GIS and Machine Learning for Small Area Classifications in Developing Countries

GIS and Machine Learning for Small Area Classifications in Developing Countries

Adegbola Ojo

CRC Press
Taylor & Francis Group
Boca Raton London New York

CRC Press is an imprint of the
Taylor & Francis Group, an **informa** business

First edition published 2021
by CRC Press
6000 Broken Sound Parkway NW, Suite 300
Boca Raton, FL 33487-2742

and by CRC Press
2 Park Square, Milton Park, Abingdon, Oxon, OX14 4RN

© 2021 Taylor & Francis Group, LLC
CRC Press is an imprint of Taylor & Francis Group, LLC

ISBN-13: 978-0-367-32244-1 (hbk)
ISBN-13: 978-0-367-65232-6 (pbk)
ISBN-13: 978-0-429-31834-4 (ebk)

DOI: 10.1201/9780429318344

To my loving parents (Mr Gabriel and Mrs Moriam Ojo)

who gave me evergreen values.

Contents

Part 2   Underlying Techniques and Deployment Approaches

Foreword

It is now nearly 50 years ago that the concept of a small area classification was first developed. Since the introduction of Potential Rating Index for ZIP Marketers (PRIZM) in the United States and A Classification of Residential Neighborhoods (ACORN) in the United Kingdom in the early 1980s, the practice of analyzing behavior by type of neighborhood has extended to over 30 countries, almost all of them in the developed world.

In most countries it was vendors of consumer products who were the first to adopt these systems or companies such as McDonalds, searching for sites accessible to the types of people who proved regular users of their restaurants, or Honda, whose models appealed to people in particular types of neighborhood. More recently, at least in Britain, these systems have also been adopted by the police, public health departments, and education authorities to target resources at communities in need of additional services.

Increasingly these tools are embedded in what are commonly known as "decision support systems," for instance, the process whereby a retailer identifies a good location for a new outlet, or a car dealer sets a sales target for each franchise.

This practice is based on a simple concept, which is as self-evident in Nigeria and Peru as it is in Sweden and Japan, that while every neighborhood is unique in certain respects, the process of urban development has resulted in most neighborhoods conforming more or less to a certain type based on its historical development and current function, with such types being broadly similar across characteristics such as housing, levels of education and income, age distribution, and household characteristics.

It is reasonable to suppose that if we can identify a set of areas that are broadly similar to each other across these different characteristics, neighborhoods belonging to any given category are likely to offer similar levels of demand for products delivered by the private sector and similar need for services delivered by the government sector. This principle provides a formal, empirical, and objective method for allocating resources, and a rational basis for allocation for arbitrating between the claims of competing political lobby groups.

From our experience in the development of these systems, it is evident that this holds just as true in the East as in the West and as true in the South as in the North. What is more is that in rapidly developing countries there are even greater disparities between types of neighborhood in terms of consumption patterns, lifestyles, and requirements for public interventions than in the countries where social classifications were first introduced. It is in countries in South America and Africa that the well-educated, the internationally connected, and the financially successful are very particular in their choice of an elite neighborhood in which to live and where conditions in rural communities vary hugely according to historical patterns of land ownership, cultivation, soil cultivation, and access to urban markets – factors typically captured by a small area classification.

What is less obvious, at least to nonstatisticians, is that the type of data that we commonly see displayed on a map are often much more revealing when analyzed using a small area classification. In other words, rather than aggregate the neighborhood information that we may hold into bigger administrative areas or regions, until we have a large enough number of cases to be statistically reliable, it makes more sense to aggregate data for neighborhoods that are broadly similar to each other, as identified by a small area classification. When data are aggregated in this way, it is much easier to see associations between different statistical measures and hence to develop hypotheses or even explanations for particular inequalities.

For example, in Britain, even 30 years after the closure of the coal-mining industry, small area classifications continue to identify communities in former mining areas as a distinctive type of neighborhood. In 2019, it was the success of the Conservative party in targeting its campaign to this very specific type of neighborhood that enabled it to win enough seats to form the government.

Over and over again, statisticians have found that the type of neighborhood one lives in is a more effective predictor of one's behavior than the actual city that one lives in.

As the practice of small area classification moves into developing countries, a number of distinct challenges and questions emerge. For example, in advanced countries, the process of associating a customer or client with a specific neighborhood typically relies on the existence of a national postcode (zipcode) system. If this does not exist, what alternative methods can link an address to a neighborhood? The author of this book recognizes this impediment alongside other data challenges and considers a number of measures for ameliorating the problem, such as the intelligent adaptation of emerging forms of data like sensors and connected devices, global positioning system (GPS) tracking data, satellite/aerial images, and nighttime visible radiation.

The rapid growth of cities in developing countries can also pose a problem. In China, for example, the rate of physical reconstruction of Shanghai and Beijing has been into such extent that any classification of neighborhoods soon gets out of date. In Brazil, by contrast, the challenge is how to attach a category to newly developing favelas.

Other than in China, with its rapid pace of physical redevelopment, the character of neighborhoods tends to change less rapidly than do their actual populations. Elite areas of Lima and Arequipa (Peru) in 2020 tend to be the same as elite areas in Lima and Arequipa in 1920, as is the case in Munich (Germany) or in Seattle (USA).

Much has been made of how the patterns of urban growth, development, and inequality in the cities of developing regions differ from that in developed regions. As the practice of small area classification spreads to developing regions, it will be interesting to find out whether the patterns of urban segregation are consistent between different nations and cities and how far these patterns depend on the extent to which public planning or market forces shape the development patterns.

This book is unique in providing both a political justification and a methodological basis for a formal analysis of urban structure in developing countries and I hope it will prove seminal in shaping an empirical understanding of how consistent these patterns are between one country and another.

Professor Richard Webber
Cofounder/Director
WebberPhillips Limited, UK
Visiting Professor
University of Newcastle, UK.

Preface

Since the emergence of contemporary area classifications, population geography has witnessed a renaissance in the area of policy-related spatial analysis. Area classifications, which subsume geodemographic classifications, often use data mining techniques and machine learning algorithms to simplify large and complex bodies of information about people and the places in which they live, work, and undertake other social activities. They are developed on the basis of geographic and social theories that suggest that people with similar characteristics are more likely to reside within the same type of locality; based on this premise, there will be different types of localities and that these types of localities will be distributed, for instance, across a nation's geographic landscape. Outputs developed from the grouping of small geographic areas on the basis of multidimensional data have proved beneficial particularly for decision-making in the commercial sectors of a vast number of countries in developed regions of the world.

One of the limitations of spatial analysis conducted in some developing countries is the difficulty in generating insight about local level discrepancies. Stakeholders within public, private, and academic sectors are increasingly in search of innovative tools and techniques for drilling down to local analytical scales for a number of reasons. Firstly, the ability to isolate developmental challenges at granular scales can help to eliminate bias and improve public confidence when disbursing national resources or deploying of socioeconomic interventions. Second, different tiers of government often need to be able to effectively target and 'market' national social policy agendas on a range of sustainable development themes. This can be particularly important when trying to educate local populations about subjects like poverty, inequality, public health epidemiology, educational disparities, and social tensions. The process requires the ability to pinpoint different local population cohorts with the correct message via the most appropriate information channel. Finally, stakeholders within developing countries need potent tools for monitoring and evaluating the local impact of national and international policies. In an age where developing countries are becoming increasingly younger and diverse, stakeholder requirements to meet these needs and others similar to them are increasingly complex and ever more important.

This book represents a chance to make the case that small area geodemographic classifications can also offer countries in developing regions of the world a distinct opportunity to address human-centered issues using novel social scientific principles and practice. The book traces the origins of small area geodemographic classifications and unmasks multiple reasons for their paucity in developing countries. Practical mechanisms for circumventing these roadblocks are explored. A template is provided for the technical development of small area classification systems that can be tailored to meet the needs of developing countries. This is followed by an array of illustrative applications and prospects for future directions.

Acknowledgements

The essential foundation of this book was established during my master of science (MSc) program in Geographic Information Science, which was completed at the University College London (UCL). During this period, I came in contact with Professor Richard Webber, originator of the postcode classification systems ACORN and Mosaic and a former director of Experian, who would later supervise my MSc dissertation. Professor Richard Webber inspired me to deepen my interest in the development and deployment of small area geodemographic tools. During my stint at UCL, I also met Dr Pablo Mateos (the Center for Research and Higher Studies in Social Anthropology, Mexico). Pablo was studying for his doctoral degree at the time and he also co-supervised my master's degree thesis. Pablo's impact and influence upon me was an important stimulant that encouraged me to pursue a doctoral degree at the University of Sheffield and he remains a lifelong friend. I am also indebted to Professor Dimitris Ballas (University of Groningen, Netherlands) and Dr Daniel Vickers (Demographics and Neighborhoods Research, UK) for their guidance and mentorship during my doctoral study at Sheffield. I also want to express my profound gratitude to Ms Irma Shagla Britton and Ms Rebecca Pringle, both are at Taylor & Francis. Their patience, support, and encouragement throughout this project have proved invaluable and I remain very grateful to them. I also wish to extend my sincere appreciation to all the reviewers of my book. Their constructive comments and suggestions helped me to sharpen my thoughts on several issues subsumed within this book.

A substantial chunk of the chapters contained in this book was written during the outbreak of the coronavirus pandemic (COVID-19) while I was observing the lockdown with my family. By implication, my wife and children had to endure weeks of diluted attention from me even though we were all on lockdown. I am forever indebted to my wife, Funmilola, and my children, Damilola, Oladapo, and Aderinsola, for their patience and understanding during this period and always. These are my heroes and I owe much to them.

Finally, although every effort has been made to ensure that the account given in this book is error-free, I recognize that it is not unlikely that there may be some omissions. The opinions expressed in the rest of this book are solely those of the author.

Adegbola Ojo
Sheffield, UK

Author Biography

Dr. Adegbola Ojo is Director of Teaching and Learning, Programme Leader, and Senior Lecturer in Urban Geography and Applications of Big Data at the School of Geography, University of Lincoln, UK. He received his PhD in Quantitative Human Geography from the University of Sheffield, his MSc in Geographic Information Science from the University College London, and his BSc in Geography and Planning Sciences from the University of Ado-Ekiti, Nigeria. His research interests are focused in understanding and representing social and spatial dynamics and intricacies of population behavior within a framework of Interdisciplinary Studies, Population Geography, Quantitative Social Science, and Computer Modeling. His research activities are grouped around the development and application of small area classifications, geographic information systems and geographic information science for informing public policy. Dr. Ojo has published many monographs and research articles with reputable journals. He has designed and delivered lectures, workshops, seminars, tutorials, practical labs, and assessments to a range of undergraduate and graduate students and working professionals.

Abbreviations

ACORN	A Classification of Residential Neighborhoods
AI	Artificial Intelligence
ANOVA	Analysis of Variance
BIC	Bayesian Information Criterion
CBD	Central Business District
CCKP	Climate Change Knowledge Portal
CCTV	Closed Circuit Television
CES	Centre for Environmental Studies
CLI	Call-line Identification
CLIQUE	Clustering in Quest
COVID-19	Coronavirus disease 2019
CPI	Corruption Perception Index
CURE	Clustering Using Representatives
CWIQ	Core Welfare Indicators Questionnaire
DA	Dissemination Areas
DBSCAN	Density-Based Spatial Clustering of Applications with Noise
DENCLUE	Density-Based Clustering
DfID	Department for International Development
DHS	Demographic and Health Survey
EA	Enumeration Area
ED	Enumeration District
ESDA	Exploratory Spatial Data Analysis
FCT	Federal Capital Territory
FOS	Federal Office of Statistics
FSI	Fragility States Index
GBD	Global Burden of Disease
GIS	Geographic Information System
GPS	Global Positioning System
GRIDCLUS	Grid-Based Hierarchical Clustering
ICT	Information and Communications Technology
IHME	Institute for Health Metrics and Evaluation
IP	Internet Protocol
IPUMS	Integrated Public Use Microdata Series-International
KDE	Kernel Density Estimation
LGA	Local Government Area
LGC	Local Government Council
MAUP	Modifiable Areal Unit Problem
MDA	Ministries, Departments and Agencies
MDG	Millennium Development Goals
MLE	Maximum Likelihood Estimation
ML	Machine Learning

NAM	Nonaligned Movement
NBS	National Bureau of Statistics
NCDC	Nigeria Centre for Disease Control
NCR	National Capital Region
NDB	National Data Bank
NGO	Nongovernmental Organization
NPC	National Population Commission
NSO	National Statistics Office
OECD	Organisation for Economic Co-operation and Development
ONS	Office for National Statistics
OPTICS	Ordering Points to Identify Clustering Structure
PAF	Principal Axis Factoring
PCA	Principal Components Analysis
PPMC	Pearson's Product Moment Correlation
PRIZM	Potential Rating Index for ZIP Marketers
RMS	Root Mean Square
SD	Standard Deviation
SDG	Sustainable Development Goals
SDI	Spatial Data Infrastructure
SIM	Subscriber Identification Module
SSA	State Statistical Agency
STING	Statistical Information Grid
UCL	University College London
UK	United Kingdom
UN	United Nations
UN DESA	United Nations Department of Economic and Social Affairs
UN IGME	United Nations Inter-agency Group for Child Mortality Estimation
UNFPA	United Nations Population Fund
UNHCR	United Nations High Commissioner for Refugees
USA	United States of America
WHO	World Health Organization

Part 1

Background, Concepts, and Definitions

1

Introduction

1.1 Global South or Developing World?

Prior to discussing some of the contemporary demographic trends and shifts, it is necessary to engage in some conceptual debates. What do we call the parts of the world that this book covers? Do we refer to these parts of the world as the Global South or developing world? Numerous development scholars continue to grapple with this important conceptual question.

Although the term "Global South" emerged in the 1950s, its first notable use dates back to 1969. Carl Preston Oglesby, an American writer, academic, and political activist, is credited with coining the term within contemporary political context. Oglesby acted as the editor of a special issue of the liberal Catholic journal called *Commonweal*. The special issue focused on the Vietnam War and Oglesby (1969) argued that centuries of US dominance over the Global South have converged to produce an intolerable social order. The founding members of the Non-Aligned Movement (NAM) also used the term politically.[1] Since the late 1960s, the term "Global South" has amassed multiple nuances and meanings (Horner and Hulme, 2019). Some have described countries within the Global South as politically and culturally marginalized (Leimgruber, 2018).

The emergence of the concept stemmed from historical attempts to define economically poorer and richer parts of the world. Many scholars and practitioners consider its predecessors less favorable while acknowledging that the term "Global South" nevertheless remains imperfect. Mahler (2017) contends that the concept of Global South subsumes at least three notions. The first notion is that it refers to nation states that are economically disadvantaged. This interpretation stems from the use within the corridors of international organizations and development partners. Mahler further

---

[1] The Non-Aligned Movement (NAM) emerged in the context of the wave of decolonization that followed World War II. The founding members of the NAM were India, Egypt, Yugoslavia, Ghana, and Indonesia.

argues that a second notion of the concept of Global South exposes the externalities of capitalism. In a sense, the concept is largely used to differentiate those parts of the world that were subjugated by others especially during the colonial and indeed the post-colonial era. The third notion of the concept of Global South is an extension of the second one. Mahler (2017) suggests that some countries interpret the concept through the lens of a shared experience of subjugation triggered by global capitalism. Those countries identify their conditions as similar and they form movements that they believe can cater to their mutual interests.

Questions remain as to the geographic limits of the Global South. Some scholars and practitioners generally use the phrase as a catchall for several countries located within four main geopolitical regions of the world: Africa, Asia, Latin America, and Oceania (Connell and Dados, 2014; Watson, 2016). For others, the literal geographic impression invoked by the term is that of the equator splitting the globe into northern and southern hemispheres. However, this is not necessarily the case because there are numerous countries within the northern hemisphere with human development and income characteristics that correspond to others within the southern hemisphere. Consequently, there is no agreed consensus in the literature regarding the designated geographic boundaries of the Global South. This works to its disadvantage. For the sake of consistency, some researchers and policymakers utilize the development and income characteristics illustrated in Figures 1.1 and 1.2, which are proposed by the World Bank and the United Nations, respectively.

It is argued that countries in the world are too diverse to be compartmentalized into two boxes: one for south and one for north. Although the term "Global South" has been increasingly popularized, it is considered too vague and much less homogenizing. The alternative term "developing world" represents a radical move from the focus on cultural or developmental dissimilarities between countries. This term gives prominence to geopolitical power relations. There are arguments that the concept should be embraced to address those spaces and peoples that have been and are being routinely negatively impacted by contemporary capitalist globalization. The term "developing world" also brings about a connotation of improving conditions, the recognition that social and economic conditions are evolutionary processes. In the remainder of this book, the term "developing world" is broadly embraced because it is considered as a more homogenizing term which also admits that countries can advance their economic growth through better utilization of natural and human resources, and that this can further result in changes in social, political, and economic structures of nations. Although the World Bank hinted in 2016 that it is retiring the term "developing world," the implication of this change is that it is likely to exacerbate global levels of inequality. There is still gross inequality within and between countries that needs to be addressed and the categorizations of developed and developing can help make the case for that.

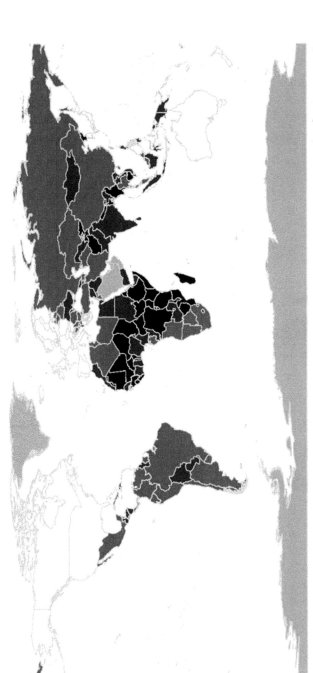

FIGURE 1.1
Map of countries by World Bank Income Groups.

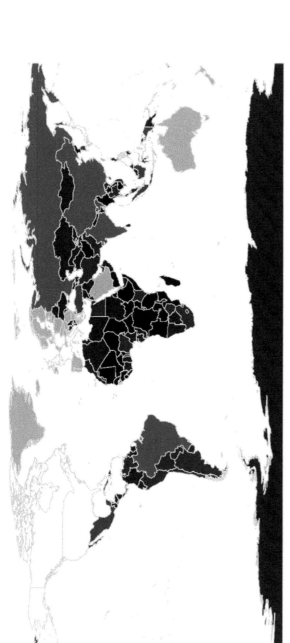

FIGURE 1.2

Map of countries by United Nations Development Groups.

1.2 Demographic Shifts across the Developing World

Drawing upon recent data released by the UN Population Division for the World Population Prospects 2019, we discovered that during the 70-year period between 1950 and 2020, the average annual global demographic growth stood at 1.60%. During this period, the demographic growth has been most pronounced within the least developed countries, which averages 2.41% annually. Similarly, low-income countries have witnessed the largest average annual population growth (2.48%). Table 1.1 summarizes the average annual change in population across six continents. Estimates shown in the table from 2020 and beyond are based on a medium fertility variant calculation. Focusing on the four continents that make up the developing world, one can discover some interesting trends. During the 150-year period, population growth will remain positive across Africa although the growth will diminish significantly to below 1% per annum from 2080 onward.

Oceania will also experience a positive population growth during this period for which figures are shown. However, the growth will diminish below 1% per annum sooner than in Africa from 2035 onward. Unlike Africa and Oceania where growth will remain positive, the populations of Asia and Latin America and the Caribbean are expected to have a decline at some point. The populations of Asia will begin to decline by 2055, while that of Latin America and the Caribbean will turn negative half a decade later – 2060.

Another story that can be told from the data presented in Table 1.1 is that the population of much of the developing world is expected to rise faster than other parts of the world especially during the 21st century. Africa will mostly contribute to this demographic trend and by the middle of the century, it is estimated that the continent would have doubled its population to about 2.5 billion people (PRB, 2016).

Different factors contribute toward the population growth. However, various studies have agreed that there are two principal reasons for the rapid population growth in much of the developing world. The first major factor accounting for the population growth across the developing world is the sharp decline in the mortality rates of infants and children (Golding et al., 2017). Considerable progress has been made in reducing under-5 mortality especially in the last 30 years. Since 1990, it is estimated that the world has reduced under-5 mortality by around 58% (Hug et al., 2018).

The second important factor that has continued to contribute to population growth in developing countries is the limited reduction in birth rates. From a global point of view, fertility levels have reached unprecedented low levels. However, stark differences persist when patterns are observed across regions. Across countries in developing continents, the average number of live births in the early 1950s was 5.5. This figure had reduced to around 2.8 live births by 2020. At present, African women have 4.4 children on average, compared to 6.6 in the early 1950s. Presently Asian women have 2.2 children

TABLE 1.1
Average annual rate of population change, 1950–2100 (percentage)

Period	Africa	Asia	Europe	Latin America & the Caribbean	North America	Oceania
1950 to 1955	2.08	1.95	0.97	2.65	1.65	2.07
1955 to 1960	2.29	1.92	0.97	2.69	1.76	2.14
1960 to 1965	2.44	2.11	0.95	2.71	1.40	2.07
1965 to 1970	2.54	2.46	0.69	2.54	1.03	2.29
1970 to 1975	2.64	2.28	0.60	2.37	0.95	1.72
1975 to 1980	2.78	1.97	0.49	2.25	0.95	1.34
1980 to 1985	2.82	1.95	0.40	2.14	0.95	1.61
1985 to 1990	2.78	1.99	0.37	1.93	0.98	1.64
1990 to 1995	2.58	1.59	0.17	1.74	1.02	1.48
1995 to 2000	2.46	1.37	−0.04	1.55	1.19	1.34
2000 to 2005	2.44	1.23	0.10	1.32	0.93	1.39
2005 to 2010	2.52	1.13	0.19	1.18	0.96	1.81
2010 to 2015	2.58	1.04	0.18	1.07	0.79	1.56
2015 to 2020	2.51	0.92	0.12	0.94	0.65	1.37
2020 to 2025	2.37	0.77	−0.05	0.84	0.59	1.21
2025 to 2030	2.25	0.62	−0.12	0.70	0.56	1.11
2030 to 2035	2.13	0.49	−0.17	0.56	0.53	1.02
2035 to 2040	2.01	0.36	−0.20	0.43	0.45	0.93
2040 to 2045	1.88	0.25	−0.23	0.32	0.38	0.86
2045 to 2050	1.74	0.14	−0.26	0.22	0.34	0.80
2050 to 2055	1.61	0.04	−0.29	0.11	0.32	0.74
2055 to 2060	1.48	−0.05	−0.33	0.02	0.33	0.69
2060 to 2065	1.36	−0.12	−0.34	−0.07	0.34	0.63
2065 to 2070	1.24	−0.19	−0.32	−0.16	0.33	0.59
2070 to 2075	1.12	−0.25	−0.28	−0.24	0.30	0.54
2075 to 2080	1.01	−0.29	−0.24	−0.30	0.27	0.50
2080 to 2085	0.90	−0.33	−0.19	−0.36	0.25	0.46
2085 to 2090	0.80	−0.35	−0.16	−0.40	0.24	0.43
2090 to 2095	0.70	−0.37	−0.14	−0.44	0.24	0.40
2095 to 2100	0.61	−0.39	−0.14	−0.46	0.25	0.37

Source: Author's elaborations based on data from the United Nations, Department of
Economic and Social Affairs, Population Division (2019). World Population Prospects
2019, Online Edition.

on average, compared to 5.8 in the early 1950s. Live births in Latin America
and the Caribbean is presently 2 per woman, compared to 5.8 in the early
1950s. For women in Oceania, it is currently estimated that each woman has
2.3 live births compared with 3.9 in the early 1950s.

As a result of comparatively higher fertility rates combined with declining
under-5 mortality, most countries across the developing world are homes
to the youngest population of the world (Weber, 2018). The huge concentration of young people in these regions often translates into higher dependency ratios as shown in Figure 1.3. A higher dependency ratio places greater
burden on the broader society as the rest of the population are required to

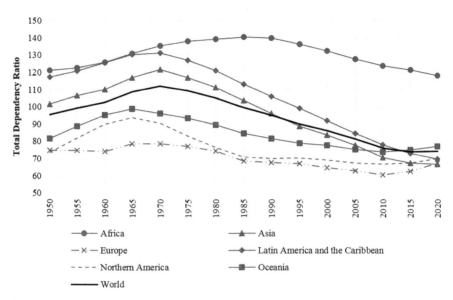

FIGURE 1.3

Ratio of population 0–19 and 65+ per 100 population 20–64, 1950–2020 (Author's elaboration based on data from the United Nations, Department of Economic and Social Affairs, Population Division (2019). World Population Prospects 2019, Online Edition).

work hard to support persons who are considered too young or too old to support themselves economically.

Future projections suggest that the working-age population will expand across much of the developing world (UN, 2015a). Some have suggested that an economic miracle can take place in the developing world if the expansion of working-age population is matched with a corresponding generation of employment opportunities (Bloom and Williamson, 1998).

1.3 The Demographic Payoff: A Myth or Reality?

The demographic payoff, also known as the demographic dividend, is a well-known concept that espouses ideas from the fields of population dynamics and development economics. It is based on the notion that demographic shifts, such as those described in the preceding section of this chapter, often lead to a transition from a largely rural and agrarian society characterized by high fertility rates to a predominantly urban industrial society with low fertility and mortality rates (Choi, 2016). Demographers have documented evidence from more developed societies that this type of transition can lead to improvements in state capacity, income, and political stability (Mason, 2005; Mason et al., 2016).

Today, the majority of nations in developing regions of the world remain predominantly youthful. An analysis of interpolated estimates from the United Nations Department of Economic and Social Affairs (UN DESA) for 2020 reveals that six in ten Africans are younger than 25 years of age. Across Latin America and the Caribbean, four in ten people are aged under 25 years. In both Asia and Oceania, three in ten people are younger than 25 years of age. Given the existence of youth bulges in many developing countries, particularly across Africa, there are debates over where and when the demographic payoff will occur. Development economists generally believe that this dividend has already arrived and that the extraordinarily large proportions of young adults in the developing countries are a source of long-term supply of ample labor, innovation, and creativity (Joe et al., 2017). It is believed that this demographic window of opportunity that numerous developing countries are currently entering offers a unique prospect for economic growth because of the size of the working-age population, which is believed to be at its maximum compared to the size of the dependent population.

Just like development economists, a number of political scientists also argue that the demographic payoff is a looming reality across the developing world (Wilson and Dyson, 2016). These political scientists believe that as political leadership improves alongside accountability, electoral reforms, increased civil liberties, and press freedom, young people will increasingly perceive themselves as important stakeholders in deciding the direction of their nations.

Unlike economists and political scientists, demographers appear less optimistic about the demographic payoff. To illustrate this pessimism, it is helpful to revisit the theory of demographic transition and the window of opportunity. Demographic transition involves changes in a country's age structure. It starts with high birth and death rates, and progresses to declining mortality, which leads to an increase in the younger age population cohorts. As this transition occurs, it creates a window of social and economic opportunity. There are at least three important features that depict the demographic window of opportunity. These include an expanding working-age population, a decreasing young cohort, and a relatively small old population cohort (Pace and Ham-Chande, 2016). Economists argue that the small population cohorts (young and old) place minimal cost pressures on society because there is a large group of working-age population that can stimulate an increase in per capita output. This rise in per capita output is what is generally described as the demographic window of opportunity (Mason, 2001; Bloom et al., 2003).

It has been suggested that the demographic window of a country opens when the median age of the population is around 26 years and ends when the median age reaches around 42 years (Valin, 2005; Wong and De Carvalho, 2006). Figures 1.4 to 1.7 show the scatterplots of the median age versus live

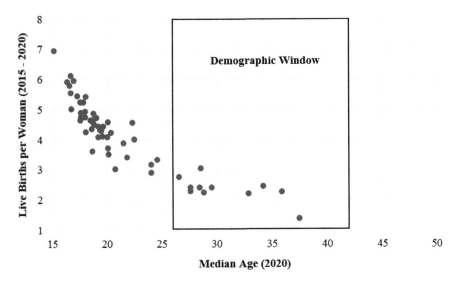

FIGURE 1.4
Scatterplot of total fertility rate (2015–2020) and median age (2020), Africa (Author's elaboration based on data from the United Nations, Department of Economic and Social Affairs, Population Division (2019). World Population Prospects 2019, Online Edition).

births for Africa, Asia, Latin America and the Caribbean, and Oceania, respectively. The analysis suggests that only 19% of African countries have reached the demographic window, with an average fertility rate of 2.4 births per woman. In Asia, 73% of countries are within the demographic window, with an average fertility rate of 2.2 births per woman. Using the median age range of 26 to 42 years, four Asian countries (China, Taiwan Province of China; Republic of Korea; China, Hong Kong; and Japan) have transited beyond the demographic window of opportunity. The story of Latin American and Caribbean countries is a little similar to Asia. Approximately 77% of countries situated within Latin America and the Caribbean are presently within the demographic window of opportunity, with an average total fertility rate of 2.1 births per woman. The United States Virgin Islands, Guadeloupe, Puerto Rico, and Martinique have all transited beyond the demographic window of opportunity. The analysis for Oceania excludes Australia and New Zealand. It suggests that only 36% of nations within that region have entered their demographic window of opportunity, with an average fertility rate of 2.3 births per woman.

There is broad agreement that the young population of much of the developing world represent the labor force of the future (Meagher, 2016). However, as shown in Figures 1.4 to 1.7, only a handful of countries in these two regions (Africa and Oceania) of the developing world have reached their demographic window of opportunity. Without reaching this important window, countries may struggle to climb up to the global middle-income

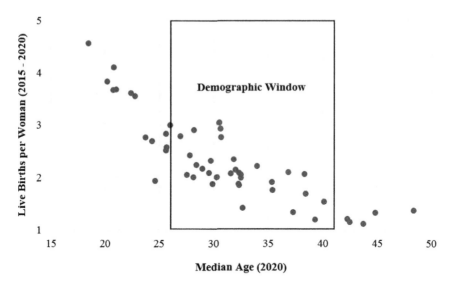

FIGURE 1.5
Scatterplot of total fertility rate (2015–2020) and median age (2020), Asia (Author's elaborations based on data from the United Nations, Department of Economic and Social Affairs, Population Division (2019). World Population Prospects 2019, Online Edition).

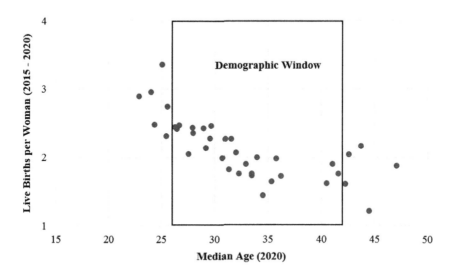

FIGURE 1.6
Scatterplot of total fertility rate (2015–2020) and median age (2020), Latin America and the Caribbean (Author's elaboration based on data from the United Nations, Department of Economic and Social Affairs, Population Division (2019). World Population Prospects 2019, Online Edition).

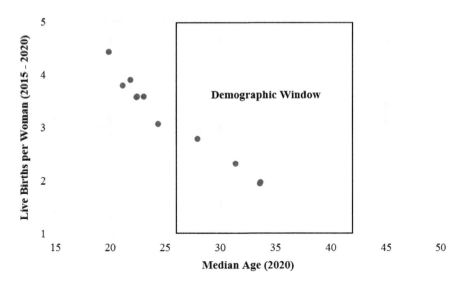

FIGURE 1.7
Scatterplot of total fertility rate (2015–2020) and median age (2020), Oceania (Author's elaboration based on data from the United Nations, Department of Economic and Social Affairs, Population Division (2019). World Population Prospects 2019, Online Edition).

category, significantly reduce under-5 mortality, and minimize late enrolment rates in secondary school.

1.4 Public Policy Challenges Arising from Shifting Demographic Patterns

During World War II, world leaders concluded that no country could be an island unto itself. They recognized the power of unity in confronting the challenges posed by the ongoing conflict, and for the first time the sitting President of the United States, President Franklin D. Roosevelt, coined the term "United Nations" in January 1942 (UN, 2008). It took another three years for the organization to be formed following ratification of a charter agreed by 5 of the 51 original member countries. As such, the United Nations (UN) formally came into existence on October 24, 1945. To date, the membership of the organization has increased to 192 member states.

The UN Charter, which serves as a building block of the UN constitution, stipulates that the rationales of the organization are to:

1 Maintain international peace and security, and to that end: to take effective collective measures for the prevention and removal of threats

to the peace, and for the suppression of acts of aggression or other breaches of the peace, and to bring about by peaceful means, and in conformity with the principles of justice and international law, adjustment or settlement of international disputes or situations which might lead to a breach of the peace;

2 Develop friendly relations among nations based on respect for the principle of equal rights and self-determination of peoples, and to take other appropriate measures to strengthen universal peace;

3 Achieve international cooperation in solving international problems of an economic, social, cultural, or humanitarian character, and in promoting and encouraging respect for human rights and for fundamental freedoms for all without distinction as to race, sex, language, or religion; and

4 Be a center for harmonizing the actions of nations in the attainment of these common ends.

A careful assessment of the key elements of the UN Charter suggests that early focus of global cooperation was dominated by a long period of cold war between some of the world's super powers. It took another five decades for the world to refocus its energy on some of the more pressing challenges confronting much of the world's population (UN, 2008) when the UN launched the Millennium Development Goals (MDGs).

Recent demographic shifts and trends experienced in developing countries have opened up multiple opportunities for residents in these countries. These opportunities include a vibrant and expanding labor force, fast-growing markets, increasing agricultural comparative advantage, urbanization, innovation diffusion, and increasing demand for the rule of law particularly among the younger population in developing countries (Ashford, 2007).

Despite the highlighted benefits of population growth, these shifting patterns also exert pressures on public policy and sustainable development. It is difficult to pigeonhole basic needs as needs vary from one person to another and from one country to another. To some priority needs are concerned with the enhancement of human capital, while others perceive the redistribution of welfare services and infrastructure as essential toward shaping the quality of life. Issues like environmental protection, energy, conflict and security, natural disaster mitigation, food security, and prevention of infectious diseases are increasingly the subjects of discussion within public policy circles (UN, 2015a).

Poverty is a multifaceted problem and large-scale concentration of population growth in developing regions continues to challenge the ability of these countries to achieve progress in reducing poverty (Ghani, 2010). Early attempts at describing this human condition focused mainly on income alone. However, the complex and dynamic nature of the phenomenon has since seen its definitions embrace concepts like insecurity, incapacitation, and inequality (Grusky et al., 2006). The multidimensional nature of poverty implies that attempts at addressing the problem in developing countries should extend beyond income poverty alone. Poverty can appear in absolute and relative forms. Absolute poverty is a situation where people live below a minimum socially acceptable standard of living, while relative poverty defines poverty in terms of a comparison of the relationship between the standards of living of different societies (Spicker, 1993; Ballas and Dorling, 2018).

Recent estimates indicate that of the world's 713 million people living in poverty (below $1.90 per day), 413 million live in Africa while 216 million are residents in South Asia (World Bank, 2018). On the basis of current trends, it can be said that nearly 9 in 10 Africans may be living in extreme poverty by 2030. It has been suggested that around 33,000 young Africans are added to those in search of jobs on a daily basis (Dieye, 2017). Among these job seekers, 60% are highly likely to remain unemployed. This inevitably exacerbates the poverty challenge confronting the continent.

Education plays a pivotal role in societal development. The processes and stages involved in educating a human being ensure that they can align their natural aptitudes and potentials to derive optimum gains. It is therefore fundamental for the human mind to think aright. Education helps to train the human mind to a point that it can think clearly in the right direction and make confident decisions. As far back as 1597, the English Philosopher and statesman Sir Francis Bacon – coined the Latin phrase *scientia potentia est*, which when paraphrased in English means "knowledge is power" (Bacon and Vickers, 2008). Over the centuries, this phrase has inspired many and helped to unearth their true underlying potentials. An educated human being has confidence to make decisions for which he or she can assume responsibilities. While it is desirable for young children and young people everywhere to receive education, it is broadly acknowledged that at least a primary level education can help people write their own destinies (Lockheed and Verspoor, 1991). As a result, the challenge of ensuring that children in developing nations have access to basic education is considered fundamental to the completeness of their growth and development. It is a fundamental human right, which is explicitly stated in the United Nations Universal Declaration of Human Rights (UN, 2015b). Unfortunately this fundamental right is still perceived by some as a mere privilege in many developing countries. Quality education remains particularly pivotal for sustaining the youth bulge in developing countries so that this shifting demographic can take part in holding their governments accountable.

Across the world, and particularly in developing countries, young women live not only in a climate of immense challenges and serious risks (Sen and Grown 1987; Peters and Wolper, 2018), but also in a world of incredible opportunity and expanding possibilities. Achieving gender parity in all areas of life is critical for balanced development. Inequality in opportunities that accrue to both men and women skews the overall development of a country. For instance, countries with some of the largest representations of marginalized women have been shown to exhibit high incidences of maternal mortality (WHO, 2005). Female education is an important tool for economic development, as women constitute the major backbone of the informal economy (Kinyanjui, 2014). The challenge of promoting equality in gender across the developing world is also linked to the participation and engagement of women in politics. This is vital because important decisions that have significant policy implications for the lives of women and children are made within political spheres. It is therefore very important for women to engage in politics and take positions especially in offices that deal directly with the affairs of women and children.

Even though fewer children under the age of five years are dying than ever before, the problem of infant mortality is still a challenge in the developing regions of the world. Based on recent estimates generated by the United Nations Inter-agency Group for Child Mortality Estimation (UN IGME) in 2018, one can discover some stark realities that make for grim reading. The data show that children born in the least developing countries are 12 times more likely to die before their fifth birthday when compared to their counterparts in North America and Europe. Preventing early childhood death in developing countries in particular remains a monumental challenge for the global health community. Although generous investments in healthcare infrastructure may help improve the situation, it has been argued that the highest returns will probably be derived from encouraging maternal immunization and breastfeeding (Girma and Berhane, 2011).

Linked to the challenge posed by children's health is the challenge of improving maternal health in fast-growing developing regions. The importance of the economic contribution of women to the growth and development of many societies in developing countries cannot be overemphasized. Boserup et al. (2007) provide a detailed review of the role of women in economic development by tracing their contributions to and challenges in rural and urban settings. When women are provided with adequate education, the scourge of poverty and related challenges can be greatly ameliorated. Not only does female education contribute to the reduction of poverty, it also emboldens women to take appropriate and timely measures for their survival and that of their children (Kelly and Kelly, 2017). Maternal mortality is connected with child delivery. Unfortunately, in many countries of the developing world where population is increasing at high rates, safe child delivery is often considered to be a privilege for those who possess economic advantage.

Corruption in some developing countries continues to undermine pro-poor development programs that should benefit the teeming population (Kieghe, 2016). There have been situations where loans granted by donor agencies or development partners are embezzled by greedy leaders rather than being deployed toward investing in the common good of the citizens of their countries (Gray and Kaufmann, 1998). Farzanegan and Witthuhn

(2017) have shown that corruption can have negative consequences on youth bulges. Their study revealed that corruption weakens political systems when the share of the youth population in the adult population exceeds around 19%. The population of young adults in most developing countries exceeds this figure.

One of the consequences of corruption and weak institutions is instability in governance (Fjelde, 2009). This contributes to fragility and civil unrest, which then lead to conflict. Civil conflicts destabilize a society and often displace people from their homes. Such an unwarranted population displacement aggravates the problem of poverty. It results in the erosion of assets or forms of livelihoods that have been amassed by the poor segments of community and replaces it with a lack of sense of dignity. The United Nations High Commissioner for Refugees (UNHCR) estimates that there are 70.8 million forcibly displaced people worldwide (UNHCR, 2019). Conflicts and displacement incapacitate the productivity of people and erode their ability to transform their aptitudes (natural and/or learning-entrenched) into meeting and sustaining their basic needs.

Unequal distribution of income and opportunities has also continued to contribute to the entrenchment of poverty in civil society across the developing world (Fosu, 2017). In many countries where governance is often aligned along ethnic or religious lines, opportunities and wealth are often constricted in the hands of very few people. Helping the poor get out of poverty therefore requires an understanding of the root causes of poverty and being able to systematically identify and target pockets of the poorest segments across the population.

An expanding proportion of young people in developing countries also presents new challenges in the determination to end some epidemics such as human immunodeficiency virus and acquired immune deficiency syndrome (HIV/AIDS). As the concentration of young people continues to increase, their risk of exposure to HIV also increases (Hogg et al., 2017). Evidence shows that although the number of new infections of the disease among children has diminished, there has not been a corresponding decline in the new infection rate among young adults (UNAIDS, 2018).

Evidence from recently published research in sub-Saharan Africa shows that natural increase is playing a prominent role in the region's fast-paced urbanization (Ojo and Ojewale, 2019). Nevertheless, rapid migration continues to play an influential role in the urbanization process in fast-growing developing economies. As masses of young people move to cities in search of employment opportunities, they place significant pressure on public infrastructure. The population growth of cities across much of the developing world has not been matched with a corresponding investment in infrastructure. This gap creates ripe opportunities for underworld activities.

Changes in the demographic structure of developing countries contribute to environmental stress in various ways. Some scholars have argued that population growth and youth bulges lead to degradation of forests, water supplies, and arable land (Kahl, 1998; Jayne et al., 2014). Long-term degradation, such as those seen in the Niger Delta of Nigeria or parts of the Horn of Africa, can stimulate conflicts, trigger rebellion, and serve as an impetus for mass migration.

Demographic changes in the developing world also translate into heightened expectations among young vibrant job seekers, particularly those who have migrated from rural to urban centers. There is the potential for civil unrest where an abundance of labor meets a shortage of jobs (Amarasuriya, 2009).

1.5 Critical Review of Global and National Policy Responses

While there are different perspectives to addressing and responding to developmental challenges posed by shifting demographic patterns, there is a general consensus that the purpose of development is to improve people's lives by expanding their choices, freedom, and dignity. In order to champion and accelerate human development programs in developing regions, at least two major global policy programs have been adopted and implemented since the turn of the twenty-first century. These include the MDGs and the Sustainable Development Goals (SDGs)

The MDGs were established following the adoption of the UN Millennium Declaration in September 2000 at the largest ever gathering of states, where it was made clear that countries, both rich and poor, should commit to do all they can to eradicate poverty, promote human dignity and equality, and achieve peace, democracy, and environmental stability (UN, 2015c). The eight MDGs included those dedicated to eradicating extreme hunger and poverty; achieving universal primary education; promoting gender equality and empowering women; reducing child mortality; improving maternal health; combating HIV/AIDS, malaria, and other diseases; ensuring environmental sustainability; and developing a global partnership for development. The global community sought to attain these goals during the period 2000 to 2015. While some scholars have argued in favor of the success of the MDGs (Abbott et al., 2017; McArthur and Rasmussen, 2017), others have been less optimistic (Clemens et al., 2007; Oleribe and Taylor-Robinson, 2016).

Following the implementation phase of the MDGs, the SDGs were adopted as the centerpiece of the 2030 agenda. The SDGs are expected to drive the global development itinerary from 2015 to 2030. The 17 SDGs are in principle an extension of the 8 MDGs, which helped to galvanize a global campaign to put an end to poverty. While the MDGs were principally dedicated toward developing countries, the SDGs are offered as a universal and more comprehensive response to the global sustainable development challenges (Loewe, 2012). The 17 SDGs, which are well publicized, are presented in

TABLE 1.2
The 17 sustainable development goals (SDGs)

Number	Goal
Goal 1	End poverty in all its forms everywhere
Goal 2	End hunger, achieve food security and improved nutrition and promote sustainable agriculture
Goal 3	Ensure healthy lives and promote well-being for all at all ages
Goal 4	Ensure inclusive and equitable quality education and promote lifelong learning opportunities for all
Goal 5	Achieve gender equality and empower all women and girls
Goal 6	Ensure availability and sustainable management of water and sanitation for all
Goal 7	Ensure access to affordable, reliable, sustainable, and modern energy for all
Goal 8	Promote sustained, inclusive, and sustainable economic growth, full and productive employment and decent work for all
Goal 9	Build resilient infrastructure, promote inclusive and sustainable industrialization, and foster innovation
Goal 10	Reduce inequality within and among countries
Goal 11	Make cities and human settlements inclusive, safe, resilient, and sustainable
Goal 12	Ensure sustainable consumption and production patterns
Goal 13	Take urgent action to combat climate change and its impacts
Goal 14	Conserve and sustainably use the oceans, seas, and marine resources for sustainable development
Goal 15	Protect, restore, and promote sustainable use of terrestrial ecosystems, sustainably manage forests, combat desertification, and halt and reverse land degradation and halt biodiversity loss
Goal 16	Promote peaceful and inclusive societies for sustainable development, provide access to justice for all, and build effective, accountable, and inclusive institutions at all levels
Goal 17	Strengthen the means of implementation and revitalize the global partnership for sustainable development

Table 1.2. A closer evaluation of the SDGs suggests that these 17 goals may be organized under a range of pillars. The pillars could include hunger (Goals 1, 2, and 6); health and wellness (Goals 3 and 6); environment (Goals 7, 11, 12, 13, 14, and 15); education and innovation (Goals 4, 8, 9, and 11); community improvement (Goals 1, 2, 4, 6, 8, 9, 10, 11, and 12); human rights and equality (Goals 1, 2, 5, 10, and 16); and international development (Goals 1, 2, 3, 4, 5, 6, 8, 9, and 10).

Since the beginning of the twenty-first century, global and national development agendas have been defined by the MDGs and subsequently by the SDGs. The effectiveness of these policy agendas has been subjected to considerable debate. Advocates of the MDGs and SDGs believe that these agendas have helped to spearhead an unprecedented global movement against extreme poverty. However, the pursuit of these global development agendas also faces fierce criticisms. They risk simplifying the concept of development by restricting the goals to what is measurable. Multiple aspects of development cannot be easily measured. Some of the goals are considered modest and some of the targets do not address problems holistically.

Much of the work of the UN and other partner agencies tends to be led by development economists. Consequently, the pursuit of economic growth dominates the programing and implementation of development agendas. These international partner agencies regularly talk about a *shared vision of development*. This phrase (shared vision of development) is often used vaguely without clear definitions and interpretations. What does development mean within the context of the MDGs or SDGs? Does it mean economic growth or does it mean human capital improvement? Development cannot be easily measured and this poses a problem for developing countries.

Another criticism levied against current policy responses is that there is quite a lot of focus on macro-level economic expansion. Evidence shows that developing countries often struggle to translate macro-economic expansion into better quality of life for people at the grassroots level (Qin et al., 2009). Focus on macro-level development has contributed toward neglecting inequalities of opportunities for different social groups within countries. More attention needs to be given to the extent to which important factors like sociodemographics and lifestyle-related issues influence within-country disadvantaged groups at micro-level.

Another shortcoming of the MDGs and SDGs is that many of the targets are rigid. These policy responses encourage problem-solving approaches that often fail to address the need for flexibility in the priorities of developing countries. For instance, the MDG on education was particularly interested in a full course of primary-level schooling but failed to address any issues on secondary and postsecondary education. This inflexible approach often promoted by some development agencies drives the programs and activities of national governments (especially in the developing world) when planning and distributing resources. This approach fails to recognize that the types and magnitudes of needs of people and communities vary from one place to another and from time to time.

Global responses also need to account more for national differences in the definitions of critical concepts and policy issues. For instance, different countries have different primary school age enrolment systems. A global development indicator might evaluate the proportion of pupils starting grade 1 who reach last grade of primary school. If one compares a country where the expected age at last grade is 11 years to a country where the expected age at last grade is 12 years, then one may find that the former would do better than the latter because students are probably likely to drop out or even die as they grow older (Sabates et al., 2010).

A further problem linked to inter-country differences in definition of terms is that data gathering mechanisms also differ along the same pathways. Countries collect data to match the definitions of indicators within their jurisdictions. Unfortunately, some statistical agencies in developing countries are not very transparent with these issues.

The reluctance of the UN and partner agencies to use their influence as mechanism for encouraging the advancement of problem-solving at the local level is somewhat surprising. The works of most agencies concentrate on reporting geographic differences at higher levels, such as regions and states. This is partly because it is generally more difficult to uncover disparities in policy issues such as health, education, and poverty at the local scale. However, it is at these local scales that the problems can quickly become endemic. Furthermore, neighborhood and community ties are generally stronger at local scales of geography (Woolcock and Narayan, 2000). It is at the local level that people feel the implications of policies designed for them. Although the impacts of strategies and policies targeted toward reducing inequalities may not quickly become apparent at the local level, they often turn out on the long run to be more effective and sustaining. This is why there is a fundamental need for a radical shift in the way the development agencies evaluate progress toward meeting the development goals. Country reports and analysis conducted at regional or state levels alone often obscure variations at the local levels.

As a result of weaknesses in addressing policy problems at the local level, the job of civil society institutions and voluntary organizations in mobilizing pressure for the attainment of these development goals is diminished. As a consequence, the success of development policy agendas is not being experienced equally across the globe. This in itself represents a major setback.

Despite the shortcomings of these development agendas and the problem-solving approaches being deployed by development agencies, there are good reasons as to why the agendas should be engaged. Firstly, the agendas help to pull together issues that demand priority attention in order to address the development debate. Secondly, these policy agendas have received tremendous support from governments around the world and can serve as the vehicle for reaching the world's least advantaged communities. Thirdly, they have the advantage of adding urgency and transparency to international development. Finally, as explicit resource commitments have been made to achieve these goals, demographers and spatial scientists need to use these to create entry points for engaging governments in development issues. Reaching and monitoring local populations has remained the missing link. The ideas, techniques, and solutions provided in this book therefore offer links between local and international actions toward human-centered development particularly in developing countries.

1.6 Why This Book Was Written

Understanding the present and future demographic shifts across the developing regions of the world can help inform development planning and policymaking. In addition, such insights can also help with identifying areas where development policy programs should be scaled to reach the teeming numbers of people in need. The clarification of the myths and realities surrounding shifting demographic profiles also offers opportunities to accelerate progress in multiple areas of sustainable development. This chapter has dissected some of the key demographic patterns and transitions that have knock-on effects on the social, economic, environmental, and political sustainability of developing regions together with the kind of public policy questions that they invoke. The declaration of the MDGs and subsequent SDGs is clearly a stated intent to urgent action. However, whether the future of developing nations is promising or perilous depends on how the international community moves to give special priority to those development goals that will give these regions a competitive edge through their youths.

The ability to translate global initiatives into local action requires the building of linkages between locally grounded action, national policies, and international priorities. It starts with a better understanding of what is happening within and across local communities based on shifting demographic patterns. It also requires the determination of and commitment from key stakeholders to revisit current approaches used for understanding, explaining, and responding to geographic inequalities at the local scale. These competing demands remain grounded in processes of socio-spatial differentiation at intra-urban and intra-rural measurement scales. For instance, the provision of educational, health, and security services in developing nations involves the disbursement of enormous budgets at the local level.

Geographic information system (GIS) is a technology that is used to integrate, analyze, and display geographic information. These technologies are becoming widespread and their capability to represent and manage relevant human development information relating to supply, demand, and milieu is increasingly making them the medium of choice for efficient and effective delivery of public policy and services in developing countries. Since the emergence of contemporary GIS, human population geography has witnessed a renaissance in the area of policy-related spatial analysis. For instance, the grouping, classification, and mapping of small administrative geographic areas is an area that has generated increasing interest among researchers of spatial sciences lately. These types of classifications are developed based on geographic and social theories that suggest that people with similar characteristics are more likely to reside within the same type of locality; based on this premise, there will be different types of localities and that these types of localities will be distributed, for instance, across a nation's geographic landscape.

Outputs developed from the grouping of small geographic areas based on multidimensional data have proved beneficial particularly for decision-making in the commercial sectors of a vast number of countries in the northern hemisphere. This book argues that small-area classifications offer countries in developing regions a distinct opportunity to address human population policy-related challenges in novel ways using area-based initiatives and evidence-based methods. A key limitation of spatial analysis conducted in numerous developing countries is the difficulty of generating insights about local area disparities when trying to understand the dynamic characteristics of human population. Stakeholders within public, private, and academic sectors are increasingly in search of innovative tools and techniques for drilling down to local analytical scales for a number of reasons. Firstly, the ability to isolate developmental challenges at granular scales can help eliminate bias and improve public confidence when disbursing national resources or deploying socioeconomic interventions. Secondly, different tiers of government often need to be able to effectively target and market national social policy agendas on a range of sustainable development themes. This can be particularly important when trying to educate local populations about subjects like poverty, inequality, public health epidemiology, educational disparities, and social tensions. The process requires the ability to pinpoint different local population cohorts with the correct message via the most appropriate information channel. Finally, stakeholders within developing countries need potent tools for monitoring and evaluating the local impact of national and international policies. In an age where many countries in the developing world are becoming increasingly younger and diverse, stakeholder requirements to meet these needs and others similar to them are increasingly complex and ever more important.

This book traces the origins of these systems and unmasks multiple reasons for their paucity in developing countries. Practical mechanisms for circumventing these roadblocks are explored. A template is provided for the technical development of small-area classifications that can be tailored to meet the needs of developing countries. This is followed by an array of illustrative applications and prospects for future directions.

References

Abbott, P., Sapsford, R. and Binagwaho, A. (2017). Learning from Success: How Rwanda Achieved the Millennium Development Goals for Health. *World Development*, 92, 103–116.

Amarasuriya, H., Gündüz, C. and Mayer, M. (2009). *Rethinking the Nexus between Youth, Unemployment and Conflict – Perspectives from Sri Lanka*. London: International Alert.

Ashford, L.S. (2007). *Africa's Youthful Population: Risk or Opportunity*. Washington, DC: Population Reference Bureau.

Bacon, F. and Vickers, B. (2008). *Francis Bacon: The Major Works*. Oxford: Oxford University Press.

Ballas, D. and Dorling, D. (2018). Spatial Divisions of Poverty and Wealth. In: D. DeBats, I. Gregory and D. Lafreniere (Eds), *The Routledge Companion to Spatial History*. Abingdon: Routledge.

Bloom, D.E., Canning, D. and Sevilla, J. (2003). *The Demographic Dividend: A New Perspective on the Economic Consequences of Population Change.* Santa Monica, CA: Rand.

Bloom, D.E. and Williamson, J.G. (1998). Demographic Transitions and Economic Miracles in Emerging Asia. *The World Bank Economic Review*, 12(3), 340–375.

Boserup, E., Kanji, N., Tan, S.F. and Toulmin, C. (2007). *Woman's Role in Economic Development.* London: Earthscan.

Choi, Y. (2016). Demographic Transition in Sub-Saharan Africa: Implications for Demographic Dividend. In: R. Pace and R. Ham-Chande (Eds), *Demographic Dividends: Emerging Challenges and Policy Implications.* Cham: Springer.

Clemens, M.A., Kenny, C.J. and Moss, T.J. (2007). The Trouble with the MDGs: Confronting Expectations of Aid and Development Success. *World Development*, 35(5), 735–751.

Connell, R. and Dados, N. (2014). Where in the World Does Neoliberalism Come from? *Theory and Society*, 43(2), 117–138.

Dieye, A.M. (2017). *How Do We Get to 50 Million Jobs by 2020! 20 July.* Kigali Convention Centre, Kigali: Rwanda.

Farzanegan, M.R. and Witthuhn, S. (2017). Corruption and Political Stability: Does the Youth Bulge Matter? *European Journal of Political Economy*, 49, 47–70.

Fjelde, H. (2009). Buying Peace? Oil Wealth, Corruption and Civil War, 1985–99. *Journal of Peace Research*, 46(2), 199–218.

Fosu, A.K. (2017). Growth, Inequality, and Poverty Reduction in Developing Countries: Recent Global Evidence. *Research in Economics*, 71(2), 306–336.

Ghani, E. (Ed). (2010). *The Poor Half Billion in South Asia*: What Is Holding Back Lagging Regions? Cambridge: Cambridge University Press.

Girma, B. and Berhane, Y. (2011). Children who were Vaccinated, Breast Fed and from Low Parity Mothers Live Longer: A Community Based Case-Control Study in Jimma, Ethiopia. *BMC Public Health*, 11, 197. Available at: https://doi.org/10.1186/1471-2458-11-197

Golding, N., Burstein, R., Longbottom, J., Browne, A.J., Fullman, N., Osgood-Zimmerman, A., Earl, L., Bhatt, S., Cameron, E., Casey, D.C., Dwyer-Lindgren, L., Farag, T.H., Flaxman, A.D., Fraser, M.S., Gething, W., Gibson, H.S., Graetz, N., Krause, L.K., Kulikoff, X.R., Lim, S.S., Mappin, B., Morozoff, C., Reiner Jr, R.C., Sligar, A., Smith, D.L., Wang, H., Weiss, D.J., Murray, C.J.L., Moyes, C.L. and Hay, S.I. (2017). Mapping under-5 and Neonatal Mortality in Africa, 2000–15: A Baseline Analysis for the Sustainable Development Goals. *The Lancet*, 390(10108), 2171–2182.

Gray, C.W. and Kaufmann, D. (1998). *Corruption and Development.* Washington, DC: World Bank.

Grusky, D.B., Kanbur, S.M.R. and Sen, A.K. (Eds). (2006). *Poverty and Inequality.* Stanford, CA: Stanford University Press.

Hogg R., Nkala B., Dietrich J., Collins A., Closson K., Cui Z., Kanters, S., Chia, J., Barhafuma, B., Palmer, A., Kaida, A., Gray, G. and Miller, C. (2017). Conspiracy Beliefs and Knowledge about HIV Origins among Adolescents in Soweto, South Africa. *PLoS ONE*, 12(2), e0165087.

Horner, R. and Hulme, D. (2019). From International to Global Development: New Geographies of 21st Century. *Development. Development and Change*, 50(2), 347–378.

Hug, L., Sharrow, D., Zhong K. and You, D. (2018). *Levels and Trends in Child Mortality.* New York, NY: United Nations Children's Fund.

Jayne, T.S., Chamberlin, J. and Headey, D.D. (2014). Land Pressures, the Evolution of Farming Systems, and Development Strategies in Africa: A Synthesis. *Food Policy*, 48, 1–17. Available at: https://doi.org/10.1016/j.foodpol.2014.05.014

Joe, W., Dash, A.K. and Agrawal, P. (2017). Demographic Dividend and Economic Growth in India and China. In: P. Agrawal (Ed), *Sustaining High Growth in India*. Cambridge: Cambridge University Press.

Kahl, C.H. (1998). Population Growth, Environmental Degradation, and State-Sponsored Violence: The Case of Kenya, 1991–93. *International Security*, 23(2), 80–119.

Kelly, D.H. and Kelly, G.P. (2017). *Women's Education in the Third World: An Annotated Bibliography*. London: Routledge.

Kieghe, D. (2016). *National Ambition: Reconstructing Nigeria*. London: New Generation Publishing.

Kinyanjui, M.N. (2014). *Women and the Informal Economy in Urban Africa: From the Margins to the Centre*. London: Zed Books.

Leimgruber, W. (2018). *Between Global and Local: Marginality and Marginal Regions in the Context of Globalization and Deregulation*. Abingdon: Routledge.

Lockheed, M.E. and Verspoor, A.M. (1991). *Improving Primary Education in Developing Countries*. New York, NY: Oxford University Press.

Loewe, M. (2012). *Post 2015: How to Reconcile the Millennium Development Goals (MDGs) and the Sustainable Development Goals (SDGs)?* German Development Institute (DIE) Briefing Paper 18/2012.

Mahler, A.G. (2017). Global South. In: E. O'Brien (Ed), *Oxford Bibliographies in Literary and Critical Theory*. New York, NY: Oxford University Press.

Mason, A. (2001). *Population Change and Economic Development in East Asia: Challenges Met, Opportunities Seized*. Stanford, CA: Stanford University Press.

Mason, A. (2005). *Demographic Transition and Demographic Dividends in Developed and Developing Countries*. United Nations Expert Group Meeting on Social and Economic Implications of Changing Population Age Structure, Mexico, UN/POP/PD/2005/2.

Mason, A., Lee, R. and Jiang, J.X. (2016). Demographic Dividends, Human Capital, and Saving. *The Journal of the Economics of Ageing*, 7, 106–122.

McArthur, J. and Rasmussen, K. (2017). *How Successful Were the Millennium Development Goals?* Brookings, 11 January. Available at: www.brookings.edu/blog/future-development/2017/01/11/how-successful-were-the-millennium-development-goals/ (Accessed: 20 March 2019).

Meagher, K. (2016). The Scramble for Africans: Demography, Globalization and Africa's Informal Labor Markets. *The Journal of Development Studies*, 52(4), 483–497.

Oglesby, C. (1969). Vietnamism Has Failed. The Revolution Can Only be Mauled, Not Defeated. *Commonweal*, 90.

Ojo, A. and Ojewale, O. (2019). *Urbanisation and Crime in Nigeria*. Cham: Palgrave Macmillan.

Oleribe, O.O. and Taylor-Robinson, S.D. (2016). Before Sustainable Development Goals (SDG): Why Nigeria Failed to Achieve the Millennium Development Goals (MDGs). *The Pan African Medical Journal*, 24, 156. doi:10.11604/pamj.2016.24.156.8447.

Pace, R. and Ham-Chande, R. (Eds). (2016). *Demographic Dividends: Emerging Challenges and Policy Implications*. Cham: Springer.

Peters, J. and Wolper, A. (Eds). (2018). *Women's Rights, Human Rights: International Feminist Perspectives*. New York, NY: Routledge.

PRB (2016). *2016 World Population Data Sheet: With a Special Focus on Human Needs and Sustainable Resource*. Washington, DC: Population Reference Bureau.

Qin, D., Cagas, M.A., Ducanes, G., He, X., Liu, R. and Liu, S. (2009). *Journal of Policy Modeling*, 31(1), 69–86.

Sabates, R., Akyeampong, K., Westbrook, J. and Hunt, F. (2010). *School Dropout: Patterns, Causes, Changes and Policies*. Paris: United Nations Educational, Scientific and Cultural Organisation (UNESCO).

Sen, G. and Grown C. (1988). *Development, Crises and Alternative Visions: Third World Women's Perspectives*. London: Earthscan.

Spicker, P. (1993). *Poverty and Social Security: Concepts and Principles*. London: Routledge.

UN (2008). *The United Nations Today*. New York, NY: United Nations.

UN (2015a). *Population 2030: Demographic Challenges and Opportunities for Sustainable Development Planning*. New York, NY: United Nations.

UN (2015b). *Universal Declaration of Human Rights*. New York, NY: United Nations.

UN (2015c). *The Millennium Development Goals Report 2015*. New York, NY: United Nations.

UNAIDS (2018). *Youth and HIV: Mainstreaming a Three-Lens Approach to Youth Participation*. Geneva: United Nations Program on HIV/AIDS.

UNHCR (2019). *Global Trends: Forced Displacement in 2018*. Geneva: United Nations High Commissioner for Refugees.

Vallin, J. (2005). The Demographic Window: An Opportunity to be Seized. *Asian Population Studies*, 1(2), 149–167.

Watson, V. (2016). Shifting Approaches to Planning Theory: Global North and South. *Urban Planning*, 1(4), 32–41.

Weber, H. (2019). Age Structure and Political Violence: A Re-Assessment of the "Youth Bulge" Hypothesis. *International Interactions*, 45(1), 80–112. WHO (2005). *The World Health Report: Make Every Mother and Child Count*. Geneva: World Health Organization.

Wilson, B. and Dyson, B. (2016). Democracy and the Demographic Transition. *Democratization*, 24(4), 594–612.

Wong, L.R. and De Carvalho, J.A.M. (2006). Age-Structural Transition in Brazil. Demographic Bonuses and Emerging Challenges. In: I. Pool, L.R. Wong and E. Vilquin (Eds), *Age-Structural Transitions: Challenges for Development*. Paris: Committee for International Cooperation in National Research in Demography (CICRED).

Woolcock, M. and Narayan, D. (2000). Social Capital: Implications for Development Theory, Research, and Policy. *The World Bank Research Observer*, 15(2), 225–249.

World Bank (2018). *Poverty and Shared Prosperity 2018: Piecing Together the Poverty Puzzle*. Washington, DC: International Bank for Reconstruction and Development/The World Bank.

2

Origins and Concept of Social Area Classification

2.1 Conceptual Clarifications

Many concepts within the social sciences are characterized by broad definitions. Social area classification, which is the centerpiece of this book, is connected with the process of segmenting geographic units into groups based on the socioeconomic and demographic features of the resident human population (Vickers and Rees, 2006). The classification of areas, which in simple terms bridges social and geographic divides (Burrows and Gane, 2006), tries to identify how geo-social similarities and dissimilarities can be better distinguished.

Geodemographics is a popular concept that has emerged within the literature on social area classification. Since its origin, various scholars have defined the concept of geodemographics in various ways. Sleight (1997) described it as the study of people by where they live. Brown (1991) particularly stressed that geodemographics concerns itself with area-based typologies and as a result proved to be adequate discriminators of the behavior of consumers. However, geodemographic theory, development, and applications transcend beyond commercial consumer and commercial use, thereby making it even more complicated. Voas and Williamson (2001) explained why it may be difficult to agree on a definition of geodemographics. According to them, in order to claim that a small geographic area is different from another, it is imperative to first clarify what is meant by *different* because of the uniqueness of these areas if their inhabitants are examined using one quality at a time.

Geodemographic classifications are types of area classifications, which attempt to simplify a large and complex body of information about people and the places in which they live, work, and undertake other social activities. They are developed based on geographic and social theories. These theories suggest that people with similar characteristics are more likely to reside within the same type of locality; based on this premise, there will be different types of localities and that these types of localities will be scattered/ distributed, for instance, across a nation's geographic landscape.

Geodemographics research is pegged upon the notion that people cluster together because of similar characteristics. This view was corroborated by Foley (1997, p. 6) who suggested that "the theory of geodemographics is based on the fact that similar people tend to cluster together and households in the same postcode sector or enumeration district can be placed in the same category". This statement can be linked to a popular hypothesis proposed by Waldo Rudolph Tobler, an American–Swiss geographer and cartographer. According to Tobler (1970, p. 236), "everything is related to everything else, but near things are more related than distant things." Tobler used this hypothesis to justify heuristic calculations in an urban growth simulation and the hypothesis eventually became known as the first law of geography. Vickers (2006, p. 16) extended Tobler's (1970) law to fit the field of geodemographics when he coined the following statement: "People who live in the same neighborhood are more similar than those who live in a different neighborhood, but they may be just as similar to people in another neighborhood in a different place."

Geodemographics is also a field that assumes that there is a linkage between the socioeconomic landscape and the conglomeration and bunching up of groups of human behaviors. Human beings have different needs that shape their living, working, and recreational choices. A better understanding of the complexities of human decisional nature can also be used to describe the communities in which people live. Such an understanding can help to tell stories about what the community is like, the other type of residents that live there, and the ways in which they are likely to behave (Harris et al., 2005).

There is a natural tendency for the human mind to structure things into groups. This helps to simplify the understanding of how the world works. The capacity to assimilate quickly and transform information into knowledge is dependent on the simplicity of presentation. If objects are considered different because of their attributes, a mass of too many objects is generated. If certain similarities can be identified among this mass of many objects, a process of ordering is determined and this gives rise to groups of similar objects. In addition to minimizing confusion and enhancing understanding, the introduction of a process of ordering units within a complex system, based on the characteristics of the units, reveals hidden patterns that exist within similar groups and between dissimilar groups (Brown et al., 2000). Everitt et al. (2011) describe the process of ordering units within social systems as the clustering process and this typically results in a classification scheme.

2.2 Pre-1980s History of Area Classifications

The origin of area classifications is generally credited to a survey conducted during the 19th century by Charles Booth, an English social reformer. This was an unprecedented enquiry of the social and economic conditions of the people of London (Orford et al., 2002). For the first time, large detail maps showing the social class of London at street geography were produced. Bales (1991) describes the rationale behind Booth's inquiry as a complex impetus driven by a blend of political and philosophic ideologies and a sense of social obligation. Booth conducted a survey that revealed that the incidence of poverty in London was far greater than what he had imagined. From his results, he concluded that 30.7% of the city's population was actually below the poverty line (Simey and Simey, 1960).

Booth's first study on poverty sought to show that poverty incidence could be measured accurately (Orford et al., 2002). This, he expected, would eventually influence the way policy was designed and ensure that such policies met actual measured needs. Booth employed a team of researchers to help him. Most of the data used were derived from lengthy interviews with professionals who according to Orford et al. (2002) had expert knowledge and experience of working with those residing in London at the time. Among the professionals consulted, the most important for Booth were the school board visitors, whom he strongly believed had detail knowledge of social conditions and poverty (O'Day and Englander, 1993). The information gathered from the school board visitors and the notes made during the survey were used to paint pictures of the general socioeconomic conditions in which people lived.

Developments in the Chicago School of Urban Sociology characterized the next phase of developments. The group at the University of Chicago was made up of urban sociologists who worked on a number of representations of social city structures (Robson, 1971). The main assumption of the Chicago School of Urban Sociology was that qualitative approaches, especially those used in naturalistic observation, were best suited for the study of urban and social problems (Becker, 1999). This ethnographic intimacy to the data generated great richness and depth to the work within the Chicago School of Urban Sociology. Nevertheless, overdependence on qualitative approaches, to the exclusion of sound quantitative techniques, later became one of the Chicago School of Urban Sociology's greatest burdens. Some notable influencers within the Chicago School of Urban Sociology whose theories are relevant to the field of social area classification are listed with corresponding theoretical models shown in Figures 2.1 to 2.3.

- Ernest Burgess (1886 to 1966) – The concentric zone theory
- Homer Hoyt (1895 to 1984) – The sector theory
- Chauncy Harris (1914 to 2003) – The multiple nuclei theory
- Edward Ullman (1912 to 1976) – The multiple nuclei theory

The concentric zones theory is a structural example of the American central city. Ernest Burgess based his theory on 1920's Chicago. He concluded that

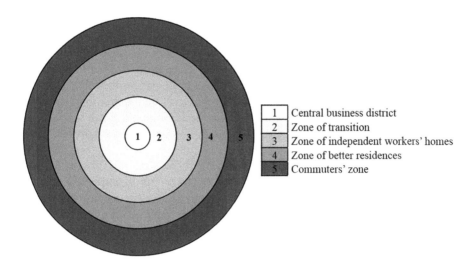

FIGURE 2.1
The concentric zone model.

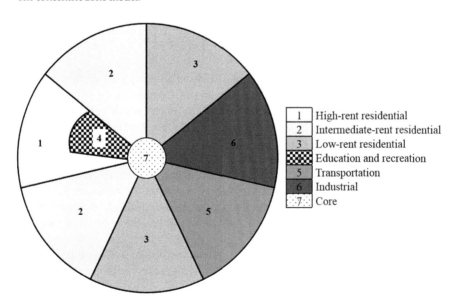

FIGURE 2.2
The sector model.

cities expand outwardly in concentric circles (Burgess, 1929). This is sometimes described as a continuous outward process of invasion or succession.
Similarly, he claimed that jobs, industry, entertainment, administrative
offices, and similar land uses were located at the center in the Central Business
District (CBD). Burgess also suggested that zone development was triggered

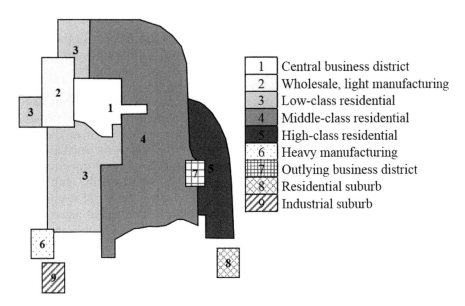

1	Central business district
2	Wholesale, light manufacturing
3	Low-class residential
4	Middle-class residential
5	High-class residential
6	Heavy manufacturing
7	Outlying business district
8	Residential suburb
9	Industrial suburb

FIGURE 2.3
The multiple nuclei model.

by competition for best location in the city. Although the concentric zone approach to classifying urban land uses is broadly appreciated in the United States of America (USA), it is less applicable outside the USA. Furthermore, the relevance of the model has decreased over time with advancements in transportation modes. It is also noteworthy to mention that in reality no strictly defined zones and boundaries exist as overlapping areas in every town. This model is also not applicable where cities are polycentric, thereby exhibiting multiple CBDs.

In the late 1930s, Homer Hoyt countered Burgess by arguing that cities do not develop in concentric circles, but in sectors (Hoyt, 1939). According to him, each sector is characterized by different economic activities. In his classificatory approach, he suggested that an entire city could be thought of as a circle and various neighborhoods within the city were modeled as sectors radiating out from the center. Hoyt's urban classification model has also faced criticism. When it was developed, only railway lines were considered for the growth of sectors; therefore, the model failed to make allowance for the use of private vehicles, which is now a prominent feature of most cities. Just like Burgess' models, the sector model is also a monocentric representation of cities failing to account for the emergence of multiple business centers. Similarly, physical or natural features may direct or restrict the growth of cities – an issue that the model does not carefully account for. Finally, the sector model does not refer to out-of-town developments.

By the mid-1940s, the mobility of people increased because of improved car ownership. This increase of movement reduced the primacy of the CBD. As the dominance of the CBD diminished, regional centers with specialized functions such as light manufacturing emerged. This was the basis of the multiple nuclei model developed by Chauncy Harris and Edward Ullman. They argued that none of the earlier models adequately reflected the changing city structure (Harris and Ullman, 1945). Their classificatory approach was particularly useful for modeling sprawling and expansive cities. Despite its strengths, the multiple nuclei model also exhibits several weaknesses. The model fails to consider the influence of physical relief and government policies. In addition, each zone displays significant heterogeneity, with little attention to homogeneity. Furthermore, some believe that it is not applicable to oriental cities with different cultural, economic, and political backgrounds.

Further developments in the USA were sparked by the publication of census small-scale data typically aimed for census tracts. The publication of this data enabled statistical methodology to be used for the first time to segment social areas in Los Angeles and San Francisco (Shevky and Williams, 1949; Shevky and Bell, 1955). More relevant research was conducted in the USA following the release of census data in 1960, covering more cities.

Another significant era in the development of area classifications in the USA was between the late 1950s and the early 1970s. Jonathan Robbin, who is credited with pioneering contemporary computer-based area classification systems (Burrows and Gane, 2006), drove the developments during this period. He combined theories from developments in the Chicago School of Urban Sociology with concepts within the sphere of factorial ecologies of positivist urban social sciences. This enabled him to produce profiles of residential Zone Improvement Plan (ZIP) code areas and formed much groundwork for the Potential Rating Index for ZIP Marketers (PRIZM). This first *modern* geodemographic system (Burrows and Gane, 2006) was based on an analysis of the groups of data from the US Census and consumer surveys.

Batey and Brown (1995) noted that in spite of the lead role set by Charles Booth in the United Kingdom (UK), nothing much was done in the UK until the 1960s. Moser and Scott (1961) employed techniques within factor analysis and were able to aggregate 157 British cities and towns into 14 groups based on four factors: social class, population change from 1931 to 1951, population change from 1951 to 1958, and overcrowding. These four factors were drawn out from 57 initial indicators, and allowed Moser and Scott to conclude on which cities were more alike. Vickers (2006) concluded that this research and the work conducted by Gittus (1964) in Merseyside and southeast of Lancashire led to a resurgence in area classification in Britain.

It is impossible to write about area classifications in the UK without acknowledging the important influence of Richard Webber. While Jonathan

Robbin added a modern touch (developing PRIZM) to area classification in the USA, the works of Webber (1977) and Webber and Craig (1976 and 1978) served as a platform for the development of modern-day geodemographics (Harris et al., 2005; Burrows and Gane, 2006; Vickers, 2006). During the 1970s, Webber, who was working with the Centre for Environmental Studies (CES), London, commenced a development of national classifications at small scales (ward, parish, and local authority levels) (Harris et al., 2005). With this development, groups of enumeration districts (EDs) could be analyzed comparatively with the national mean for certain census variables (Webber and Craig, 1976 and 1978). This development was significant as it had been previously difficult to embark on a countrywide comparison because of the different methodologies employed by different studies that had been conducted (Vickers, 2006). A company called CACI subsequently acquired the classification system developed by Webber. CACI was established in 1975 and primarily focused on providing information technology (IT) and marketing solutions for organizations in different sectors of the economy. CACI launched Webber's classification with a new name: A Classification of Residential Neighborhoods (ACORN). A subsequent linkage of the classification system to the postcode geography provided significant discriminating information about consumer behaviors in different neighborhoods for practitioners in the private sector (Harris et al., 2005). The modern-day geodemographic industry resulted from the continuous development of further versions of ACORN.

2.3 Area Classifications in the 1980s and 1990s

CACI remained the sole player within the UK geodemographics market between the 1970s and the early 1980s; however, competition from other emerging companies resulted in a gradual drift of some of CACI staff (Sleight, 2004; Harris et al., 2005). Developments in the industry in the mid-1980s were enhanced by the release of the 1981 UK Census. Richard Webber moved from CACI to CCN Marketing and the first version of the popular Mosaic classification system was launched in 1986. Webber (1977) used only the census data to develop earlier classifications. However, his work at CCN Marketing in the 1980s led to the introduction of non-census data for the development of area classifications.

The increased use of area classifications within the private sector had a number of underlying reasons. Of particular importance were significant developments, which occurred within the sphere of retailing and marketing. Beaumont and Inglis (1989) identified a transition from an emphasis on *mass marketing* to what they termed as *niche marketing*. The dominance of the industry by commercial segmentations seemed to have overshadowed academic interests in developing area classifications in the UK. The first purely academic research is attributed to Marcus Blake and Stan Openshaw, who developed a general-purpose classification system using an unsupervised neural network technique (Blake and Openshaw, 1995). Their methodology is based on Kohonen's self-organizing map. It provides an avenue whereby the number of assumptions is reduced as much as possible (Openshaw, 1984a). Their approach however left a number of areas unclassified. Openshaw and Blake (1995) utilized data from the 19th UK census, popularly called Census 1991.

2.4 Post-2000 Area Classifications

Following the release of data from Census 2001, the UK Office for National Statistics (ONS) commissioned the development of a suite of area classifications for small geographic areas in the UK (Vickers and Rees, 2006). This work that was conducted at the University of Leeds generated an openly available methodology for classifying UK Output Areas. Output Areas are the smallest spatial scale at which census data are released in the UK. The work of Vickers (2006) subsumed 41 variables, which were selected from an original 129 variables. These 41 variables were categorized under five domains: demographic, household composition, housing, socioeconomic, and employment domains (Vickers, 2006). The Output Area classification utilized an iterative relocation algorithm commonly referred to as K-means.

The exposure of Canada and the USA to area classifications in contemporary times has not been very different. Their markets have been dominated by commercial segmentations, which focus on market research activities. Following the release of data from the Canadian Census of 2001 and the addition of a new geographic unit called Dissemination Areas (DAs), a team of researchers from a company called Environics Analytics Group developed an area classification called PRIZM. PRIZM embedded selected variables from the Canadian Census and lifestyle data described as behavioral and attitudinal variables, which were derived from surveys. The rationale for including attitudinal variables was that the classification would perform better in practice revealing trends not covered in the census (Environics Analytics, 2019). One challenge with this was that the databases from which these lifestyle data were covered were only a fraction of spatial units from the national structure. In addition, the geography for sampling the behavioral and attitudinal variables is different from the geography of the census data. This meant that data for other areas were interpolated or simulated, which

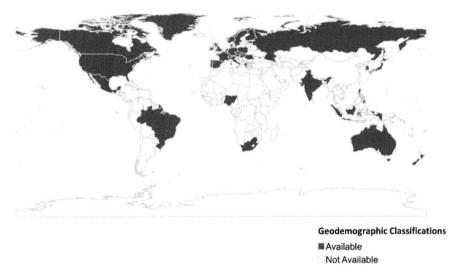

Geodemographic Classifications
■ Available
☐ Not Available

FIGURE 2.4
Spread of academic and commercial small area geodemographic classifications.

exposed the data to errors. Just like other commercial companies in the UK and the USA, the methods used to develop PRIZM have not been made publicly available to allow for academic scrutiny.

Figure 2.4 shows those countries where academic and commercial small area classifications have been built to segment populations on a geodemographic basis. The spread of small area classification approaches across developing regions of the world has been relatively slow. Most of the commercial companies that have built these classifications focus on Euro-American markets (Leventhal, 2016) where their primary aim is consumer target marketing. They add non-census data to available census statistics when building area classifications. These non-census data are also called lifestyle data, which is a term encompassing data collected based on sampled surveys of consumer choices and behaviors (Sleight 2004; Harris et al. 2005). Some scholars believe that the addition of such data to the development of area classifications helps bridge the gap often created between the conduct of successive censuses (Harris et al., 2005). Vickers (2006) has argued that in terms of quality, such data may not be properly evaluated and can be biased toward affluent sections of the society.

Since the early 2000s, there has been a push toward making public data more openly available (Molloy, 2011). This has helped to make large demographic and lifestyle data sets relatively cheap or free to access, particularly in developed countries. The combination of open data services, unprecedented computing power in GIS, web-based services, increasing market demand, and readily available capital base has underpinned the explosion of small area classifications in developed countries since the beginning of the current millennium (Singleton and Spielman, 2014).

2.5 Criticisms Levied against the Social Sorting of People in Small Geographic Areas

Similar to several other innovative developments within the social sciences, the development and use of geodemographic classifications have attracted some measure of criticisms. Much of the criticisms are connected with theoretical and methodological issues like the ecological fallacy and the modifiable areal unit problem (MAUP). There are also concerns about the longevity of the data, which act as the building blocks of these classifications. Some scholars have also raised concerns about the ethical implications of an over-coded society (Burrows and Gane, 2006). This section focuses on a review of some of these key criticisms.

2.5.1 Scale

Scale plays a crucial role when undertaking any form of geographic analysis (Marceau, 1999; Moore, 2008). The term "scale" has several meanings, depending on the context in which it is used. For instance, the scale of a map is defined as the ratio of a distance on the map to the corresponding distance on the ground. However, within the context of social area classifications, the concept of scale is used in a way similar to human geography. Scale is used here as a form of interconnected hierarchy. For instance, consider a hierarchy of local, regional, national, and global geographies. Each level in this hierarchy can be viewed as a geographic scale of analysis.

It is often challenging to secure demographic and socioeconomic data for small areas. Therefore, when developing geodemographic classifications, it is common practice to aggregate data upward or downward (Harris et al., 2005; Leventhal, 2016). Jarvis (1996) describes these processes as upscaling and downscaling data. When transferring socioeconomic data from one geographic scale to another, the geometric structure of the data and their associated attributes may be distorted (Marceau, 1999). The twin issues of the MAUP and the ecological fallacy can also help to explain the implications of this distortion.

2.5.2 True Representativeness

Another issue that confronts proponents and users of small area classifications is the problem of representativeness. Geographic modeling of the real world by way of grouping data often requires generalizations to be made about people and places. The process of generalizing information therefore triggers questions on how homogeneous the groups created from classification procedures truly are. For grouped data to be truly representative of people and places, there must be a significant level of intra-member homogeneity within the group (Kamimura, 2000). This regular problem confronts small area classifications. How representative are they of the people and places that are classified?

In geostatistical analysis, ecological fallacy occurs when the characteristics of people and places are deduced from inferences about the group to which those people and places belong (De Smith, 2007). Some critics of geodemographic classifications have suggested that an analysis of area-level data may give rise to conclusions that are different from unit-level data (Steel and Holt, 1996). While this may be true, it is important to stress that all forms of geographic analyses experience the ecological fallacy (Vickers, 2006). Steel and Holt (1996) have suggested that the key to analyzing data from grouped populations is the development of statistical models for the individuals, the groups, and the interactions between them.

2.5.3 Statistical Biasing Effect

When the same basic socioeconomic data are aggregated in different ways, as is often the case with small area classifications, they can yield different results. This is a geographic problem known as the MAUP. It is a statistical biasing effect where the same data tell conflicting stories when they are aggregated to multiple areal units, such as postal units, enumeration districts, electoral wards, local authorities or policing, and health geographies. Gehlke and Biehl (1934) first identified the problem, but the term MAUP was formally coined by Openshaw and Taylor (1979).

To help illustrate the MAUP, Taylor et al. (2003, p. 42) wrote the following about the UK census data:

> The UK census collects individual household level data and then aggregates up to a variety of larger zones, such as the Enumeration District, Ward or Local Authority. However, these zones determined for ease of enumeration may bear little resemblance to the social geography of the people they contain. Consequently, the analysis of such data in different zones, or levels, may alter the resulting pattern of aggregated observations

Openshaw and Taylor (1979) examined the effects of the MAUP in detail. They considered three experiments within different statistical and spatial contexts. They explored data on the percentage of elderly voters and correlated these with Republican Party voters. The revelations were stunning. They found that if 99 counties making up the state of Iowa were aggregated into larger and fewer districts, and a number of combinations of these districts were considered, the correlation values could range from +0.97 to −0.81. Fotheringham and Wong (1991) were also able to demonstrate that the MAUP has implications for multivariate statistical analysis. Vickers (2006) illustrated the effect of the MAUP using arbitrary census areas. From his experiments, he concluded that there is no real solution to the problem – an inference which is often drawn by many researchers who have worked on the MAUP (see Gehlke and Biehl, 1934; Openshaw, 1984b; Fotheringham and Wong, 1991; Wong, 2004; De Smith, 2007).

Some solutions have been made to help combat the effects of the MAUP. The application of clustering techniques for grouping data was one of the first approaches. The aim is to ensure that the numbers and configuration of the clusters portray as much information as possible on the phenomena of interest. Openshaw (1984b) and Larson (1986) have criticized this methodology recognizing that zoning systems change with respect to the problem being investigated.

Jelinski and Wu (1996) have also put another useful suggestion forward. They advocated for the use of individual non-modifiable entities to perform analysis. They however recognized the difficulty with this approach, noting that it is not always possible to identify the individual entities of certain measures. They also admitted that the approach may result in too much detail, which gives rise to more complex or impracticable analysis.

Another interesting proposition is to perform a sensitivity analysis to determine which variables are sensitive in terms of scale and configuration and the severity of such sensitivity (Jelinski and Wu, 1996). Fotheringham (1989) has used this method to explore and explain the rate of change of variables and relationships of interest with respect to scale.

While some researchers have regarded the MAUP in practice as not as big an issue as it might appear (De Smith, 2007), others (Openshaw and Taylor, 1981; Openshaw, 1984b) have stressed the severity of the problem.

2.5.4 Longevity of Socioeconomic Data Inputs

The main data inputs for geodemographic classifications are derived from national censuses (Harris et al., 2005). Even in developed countries where censuses are conducted frequently (within intervals of 10 years), detailed results are normally not published until about three years after data collection. In some developing countries censuses have become infrequent (Leete, 2001) and the publication of detailed results takes far longer. By the time detailed census data are released in some countries, some data users believe that the data would have become outdated. Using the UK example, Vickers (2006) recounts that about 12% of people relocate each year and 77% of people remain in their enumerated locations when results are published. He also highlighted the inevitable circumstances of deaths and births and the difficulty associated with updating census data between censuses. However, spatial microsimulation techniques are increasingly being recognized as a useful means of dealing with the problem of updating data and estimating socioeconomic indicators for small areas (Ballas et al., 2004, 2005). This works well where there is reliable baseline data including good-quality population survey microdata.

In developing countries, there are urban communities, which are characterized as slums (Davis, 2005). Slums often attract urban regeneration projects, which generally result in the relocation of people from one place to another. Urban regeneration activities change environmental, social, and economic conditions of people and places. Some believe that these urban renewal activities contribute toward distorting census data especially if renewal projects take place after the census has been conducted. To combat this problem, commercial providers of geodemographic classifications, in particular, often include non-census data in their classifications. Many of these data are associated with the lifestyles of people, which are reflected by economic factors and, to some extent, social trends. These social trends are changing even more rapidly (Butler and Savage, 2013), making it even more important to frequently update lifestyle data.

An underlying goal of geodemographic classification is the ability to describe the type of neighborhood an individual lives from the socio-economic profile of all residents in that neighborhood. This implies that these typologies can also act as surrogates for the social patterns within society. Orford et al. (2002) re-examined Charles Booth's work over a century after the survey was conducted. They produced digitized maps from Booth's poverty maps of London and were able to examine these within a GIS. In contrast with the 1991 census in the UK, their statistical enquiry revealed that absolute poverty had decreased in London over the century. However, in relative terms, London had not changed much. Orford et al. (2002) concluded that the poorest people in London remained in the same neighborhoods as at the time of Booth's survey. In most cases, people change their neighborhoods to fit their lifestyles. The extent of the aging effect of socioeconomic statistics on the development of area classifications depends on the size of the areal units (Vickers 2006). Classifications based on large areal units generally prove less susceptible to change over time.

2.5.5 Overstating Capabilities of Area Classifications

The methodological approaches used by commercial providers of small area classifications are not in the public domain. Vickers (2006) pointed out that some of the commercial companies make extraordinary claims about the quality of their products without providing opportunities for researchers to validate the claims. He further argued that by providing so much precise information about people sorted into groups, there is a tendency to misrepresent reality and ignore the diversity of communities. The early work of Charles Booth, for instance, showed that within-group variability exists among members of a group (Harris et al., 2005). It is therefore essential to provide relatively broad descriptions when commenting on characteristics of groups produced from after developing a geodemographic classification.

2.5.6 Ethical Issues

An underlying logic of geodemographics is the idea that people are whom they are because of where they live. Commercial classifications are built on the notion that lifestyles assemble. It is therefore common to hear statements such as *we know you because we know where you live*. These classifications embed information from a variety of sources, notably credit reference information, customer loyalty schemes, lifestyle databases, and increasingly social media footprints, which are primarily used for target marketing (Curry, 1998; Leventhal, 2016). There is increasing data protection, confidentiality, and civil liberty concerns with access to individual level geospatial data (Boyd and Crawford, 2012; Wilson, 2012; Crampton et al., 2013). Some commercial small area classifications are tools that are sometimes derived from interrogating databases containing personal information. Some consider the analysis of such data sets to be an infringement on privacy.

Graham (2005) considered geodemographics as an information and communications technology (ICT) approach for sorting social space. In his view, this will only provide enhanced services to those persons or neighborhoods that are deemed as attractive, while the less attractive ones will lag behind. The continuous development and provision of geodemographic profiles also gives room for people to sort themselves out (Burrows and Gane, 2006). Burrows et al. (2005) cited examples of websites in the USA where people supply their sociodemographic preferences and are provided with zip codes that match those preferences. Examples like this indicate that using the power and results of geodemographic classifications wrongly could lead to further segregation and polarization within the society.

It has also been argued that geodemographic classifications have caused some public service providers to become biased when targeting services to the public. They introduce geodemographic filters in cases where everyone could be considered on a level playing ground. Burrows and Gane (2006) argued that this is a negative implication of social space sorting. According to them

> geodemographic categories accessed via a postcode or something that can be linked to a postcode (such as a telephone number or an e-mail address) can be "invisibly" used by socio-technical systems to softwaresort places (and of course the people who live in them). This might happen in an increasing number of contexts such as call-line identification (CLI) queuing systems in (increasingly ubiquitous) telephone call-centers; the determination of insurance premiums; credit ratings (see, for example,

www.checkmyfile.com); and, possibly, selection procedures in higher
education and for employment.

(Burrows and Gane, 2006 p. 805)

Vickers (2006) described this attitude as postcode persecution, suggesting
that the labeling of certain areas may contribute to treating all individuals
associated with that area unfairly. Most of these practices are connected to
commercial applications of geodemographics. Many profit-centric commercial organizations throw caution to the wind and exploit the discriminating
power of these data-driven classifications negatively (Kitchin, 2016). There
are also no clear checks and balances in the industry to control the manner in
which these companies use the classifications. Attempts to control the preda-
tory practices of commercial companies by limiting access to existing and
new forms of socioeconomic data is proving counterproductive for more ethical users in the academic and research community.

2.6 Conclusion

Geodemographics is a concept that has emerged from the tradition of classifying geographic areas. Many academics and practitioners trace the origins of
geodemographics to Charles Booth's poverty maps of London. Developments
in the Chicago School of Urban Sociology also exerted significant influence
on the origins of geodemographics. The grouping of places into sets, which
are meaningful and simple in terms of their general population attributes,
is an important feature of social scientific investigation. Area classifications
therefore did not originate without key socioeconomic implications. It is a
field that emerged from conscious and unconscious efforts to understand
the patterns and processes within geographic and social space with a view
to solving problems, particularly those relating directly to human population. Since the days of Charles Booth, and the emergence of modern-day
geodemographics, the primary goal of these classifications has been to create
relatively homogeneous groups that reflect the dominant characteristics of
resident population. The simplicity and yet potency of these data-driven
classifications has resulted in greater interest from commercial organizations
across the more developed nations. Although these organizations have
contributed toward the growth and explosion of the industry, some of their
practices also play a major role in the criticisms levied against the development and use of geodemographic classifications. As discussed in this chapter,
the existing challenges in the field are both methodological and ethical.
Methodological challenges may be ameliorated by involving the academic
community in research and encouraging the rigorous scrutiny of techniques

used to build these classifications. Ethical shortcomings will require better and balanced legislation that does not prove counterproductive for the academic community.

References

Batey, P.W.J. and Brown, P.J.B. (1995). From Human Ecology to Customer Targeting: The Evolution of Geodemographics. In: P. Longley and G.P. Clarke (Eds), *GIS for Business and Service Planning*. New York, NY: John Wiley & Sons.

Ballas, D., Kingston, R., Stillwell, J. and Jin J. (2004). Building a Spatial Microsimulation Decision Support System. *Conference on Geographic Information Science*, Heraklion, Greece.

Ballas D., Rossiter D., Thomas B., Clarke G. and Dorling D. (2005). *Geography Matters. Simulating the Local Impacts of National Social Policies*. York: Joseph Rowntree Foundation.

Beaumont, J.R. and Inglis, K. (1989). Geodemographics in Practice: Developments in Britain and Europe. *Environment and Planning A*, 21(5), 587–604.

Becker, H.S. (1999). The Chicago School, So-Called. *Qualitative Sociology*, 22(1), 3–12.

Blake, M. and Openshaw, S. (1995). *Selecting Variables for Small Area Classification of 1991 UK Census Data*. Working Paper 95/2, School of Geography, University of Leeds. Available at: www.geog.leeds.ac.uk/papers/95-2/ (Accessed 07 July 2019).

Boyd, D. and Crawford, K. (2012). Critical Questions for Big Data. *Information, Communication and Society*, 15(5), 662–679.

Brown, P.J. (1991). *Exploring Geodemographics*. In: I. Masser and M.J. Blakemore (Eds), *Handling Geographical Information*. London: Longman.

Brown, P.J.B., Hirschfield, A.F.G. and Batey, P.W.J. (2000). *Adding Value to Census Data: Public Sector Applications of Super Profiles Geodemographic Typology*. Working Paper 56, URPERRL, Department of Civic Design, University of Liverpool.

Burgess, E.W. (1929). Urban Areas. In: T.V. Smith and L.D. White (Eds), *Chicago: An Experiment in Social Science Research*. Chicago, IL: University of Chicago Press.

Burrows, R., Ellison, N. and Woods, B. (2005). *Neighborhoods on the Net: Internet-Based Neighborhood Information Systems and Their Consequences*. Bristol: Policy Press.

Burrows, R. and Gane, N. (2006). Geodemographics, Software and Class. *Sociology*, 40(5), 793–812.

Butler, T. and Savage, M. (Eds). (2013). *Social Change and the Middle Class*. London: Routledge.

Crampton, J., Graham, M. and Poorthuis, A. (2013). Beyond the Geotag: Situating "Big Data" and Leveraging the Potential of the Geoweb. *Cartography and Geographic Information Science*, 40(2), 130–139.

Curry, M. (1998). *Digital Places: Living with Geographic Information Technologies*. London: Routledge.

Davis, M. (2005). *Planet of Slums*. London: Verso.

De Smith, M.J., Goodchild, M.F. and Longley, P.A. (2007). *Geospatial Analysis: A Comprehensive Guide to Principles, Techniques and Software Tools*. Leicester: Matador.

Environics Analytics (2019). *Technical Documentation: PRIZM 2019*. Available at: https://environicsanalytics.com/docs/default-source/can---glossaries-and-technical/prizm-2019---technical-document.pdf?sfvrsn=638227d3_14 (Accessed 27 June 2019).

Everitt, B.S., Landau, S., Leese M. and Stahl, D. (2011). *Cluster Analysis*. London: Wiley.

Fotheringham, A.S. (1989). *Scale-Independent Spatial Analysis*. In: M. Goodchild and S. Gopal (Eds), *The Accuracy of Spatial Databases*. London: Taylor & Francis.

Fotheringham, A.S. and Wong, D.W.S. (1991). The Modifiable Areal Unit Problem in Statistical Analysis. *Environment and Planning A*, 23(7), 1025–1044.

Foley, T. (1997). Business Minded. *New Perspectives*, 6, 6.

Gehlke, C.E. and Biehl, K. (1934). Certain Effects of Grouping upon the Size of the Correlation Coefficient in Census Tract Material. *Journal of the American Statistical Association Supplement*, 29, 169–170.

Gittus, E. (1964). The Structure of Urban Areas: A New Approach. *Town and Planning Review*, 35(1), 5–20.

Graham, S.D.N. (2005). Software-Sorted Geographies. *Progress in Human Geography*, 29(5), 1–19.

Harris, C.D. and Ullman, E.L. (1945). The Nature of Cities. *The ANNALS of the American Academy of Political and Social Science*, 242(1), 7–17.

Harris, R., Sleight, P. and Webber, R. (2005). *Geodemographics, GIS and Neighborhood Targeting*. London: Wiley.

Hoyt, H. (1939). *The Structure and Growth of Residential Neighborhoods in American Cities*. Washington, DC: Federal Housing Administration.

Jelinski, D.E. and Wu, J. (1996). The Modifiable Areal Unit Problem and Implications for Landscape Ecology. *Landscape Ecology*, 11(3), 129–140.

Kamimura, R.T., Bicciato, S., Shimizu, H., Alford J. and Stephanopoulos, G. (2000). Mining of Biological Data II: Assessing Data Structure and Class Homogeneity by Cluster Analysis. *Metabolic Engineering*, 2, 228–238.

Kitchin, R. (2016). The Ethics of Smart Cities and Urban Science. *Philosophical Transactions of the Royal Society A*, 374(2083), 1–15.

Larson, R.C. (1986). The Invalidity of Modifiable Areal Unit Randomization. *The Professional Geographer*, 38(4), 369–374.

Leete, R. (2001). Population and Housing Censuses: A Funding Crisis? *Symposium on Global Review of 2000 Round of Population and Housing Censuses: Mid-Decade Assessment and Future Prospects*. Statistics Division, Department of Economic and Social Affairs, United Nations Secretariat, New York, 7–10 August.

Leventhal, B. (2016). *Geodemographics for Marketers: Using Location Analysis for Research and Marketing*. London: Kogan Page.

Marceau, D.J. (1999). The Scale Issue in Social and Natural Sciences. *The Canadian Journal of Remote Sensing*, 25(4), 347–356.

Molloy, J.C. (2011). The Open Knowledge Foundation: Open Data Means Better Science. *PLoS Biol*, 9(12), e1001195.

Moore, A. (2008). Rethinking Scale as a Geographical Category: From Analysis to Practice. *Progress in Human Geography*, 32(2), 203–225.

Moser, C.A. and Scott, W. (1961). *British Towns: A Statistical Study of Their Social and Economic Differences*. Edinburgh: Oliver & Boyd Ltd.

O'Day, R. and Englander, D. (1993). *Mr. Charles Booth's Inquiry: Life and Labor of the People in London Reconsidered*. London: The Hambledon Press.

Openshaw, S. (1984a). Ecological Fallacies and the Analysis of Areal Census Data. *Environment and Planning A*, 16(1), 17–31.

Openshaw, S. (1984b). *The Modifiable Areal Unit Problem: Concepts and Techniques in Modern Geography*. Norwich: Geo Books.

Openshaw, S. and Taylor, P.J. (1979). A Million or So Correlation Coefficients: Three Experiments on the Modifiable Areal Unit Problem. In: N. Wrigley (Ed), *Statistical Applications in the Spatial Sciences*. London: Pion.

Orford, S., Dorling, D., Mitchell, R., Shaw, M. and Smith, G.D. (2002). Life and Death of the People of London: A Historical GIS of Charles Booth's Inquiry. *Health and Place*, 8(1), 25–35.

Robson, B.T. (1971). *Urban Analysis: A Study of City Structure*. Cambridge: Cambridge University Press.

Shevky, E. and Bell, W. (1955). *Social Area Analysis: Theory, Illustrative Application and Computational Procedures*. Stamford, CA: Stamford University Press.

Shevky, E. and Williams, M. (1949). *The Social Areas of Los Angeles: Analysis and Typology*. Berkeley, CA: University of California Press.

Simey, T. and Simey, M. (1960). *Charles Booth. Social Scientist*. London: Oxford University Press.

Singleton, A.D. and Spielman, S.E. (2014). The Past, Present, and Future of Geodemographic Research in the United States and United Kingdom. *The Professional Geographer*, 66(4), 558–567.

Sleight, P. (1997). *Targeting Customers: How to Use Geodemographic and Lifestyle Data in Your Business*. Henley-on-Thames: NTC Publications.

Sleight, P. (2004). *Targeting Customers: How to Use Geodemographic and Lifestyle Data in Your Business*. Henley-on-Thames: World Advertising Research Centre.

Steel, D.G. and Holt, D. (1996). Analyzing and Adjusting Aggregation Effects: The Ecological Fallacy Revisited. *International Statistical Review*, 64(1), 39–60.

Taylor, C., Gorard, S. and Fitz, J. (2003). The Modifiable Areal unit Problem: Segregation between Schools and Levels of Analysis. *International Journal of Social Research Methodology*, 6(1), 41–60.

Tobler, W.R. (1970). A Computer Movie Simulating Urban Growth in the Detroit Region. *Economic Geography*, 46, 234–240.

Vickers, D.W. (2006). *Multi-level Integrated Classifications Based on the 2001 Census*. Unpublished PhD Thesis. School of Geography, University of Leeds.

Vickers, D. and Rees, P. (2006). Introducing the Area Classification of Output Areas. *Population Trends*, 125, 15–24.

Voas, D. and Williamson, P. (2001). The Diversity of Diversity: A Critique of Geodemographic Classification. *Area*, 33(1), 63–76.

Webber, R. (1977). *An Introduction to the National Classification of Wards and Parishes*. Planning Research Applications Group Technical Paper No. 23. London: Centre for Environmental Studies.

Webber, R. and Craig, J. (1976). Which Local Authorities Are Alike? *Population Trends*, 5, 13–19.

Webber, R. and Craig, J. (1978). Socioeconomic Classifications of Local Authority Areas. *Studies in Medical and Population Subjects*, 35. London: OPCS.

Wilson, M. (2012). Location-Based Services, Conspicuous Mobility, and the Location-Aware Future. *Geoforum*, 43(6), 1266–1275.

Wong, D.W.S. (2004). *The Modifiable Areal Unit Problem (MAUP)*. In: D.G. Janelle, B. Warf and K. Hansen (Eds), *WorldMinds: Geographical Perspectives on 100 Problems*. Dordrecht: Springer.

3

Public Policy Prospects of Small Area Classifications for Developing Countries

3.1 Unmasking Subnational Population Disparities

Shifts in the demographic composition of developing countries have given rise to new challenges and opportunities. Consequently, there is increasing demand for progress in the measurement of indicators used to ascertain the obstacles confronting people as well as the progress being made toward improved well-being in developing regions (Jacob, 2017). These measures play a fundamental role for policymakers. They allow them to plan and validate the effectiveness of their policies. In order to be informative, indicators used to evaluate human progress and development need to be constructed and used at the appropriate geographic scale of disaggregation (Liverman, 2018).

There is increasing demand for understanding human development issues at subnational geographies (Macfarlane et al., 2019). To gratify the rising demand of statistical estimates on human population progress and differences at subnational scales, there is the need to resort to small area methodologies. Small area methodologies help to fill the gap between official statistics and local request of data.

One of the challenges that confront spatial scientists in developing countries is the problem of generating insights into local-level population discrepancies. When trying to gauge intra-country population disparities, a substantial number of developing countries depend on estimates generated by agencies like the World Bank and the UN. Unfortunately, most of these agencies restrict their estimates and analysis to higher-level geographies such as regions and states. There is minimal impetus from international donors and agencies to encourage national statistical agencies in developing countries to generate development indicators at local-level subnational geographies.

Similarly, wide-ranging academic research practices in developing countries are often restricted to higher-level geographies (Ojo and

Ezepue, 2011). Profound human development topics like poverty and well-being are often analyzed and reported using larger geographic units. Variations at subnational levels can often be overlooked despite the fact that differences in outcomes and distribution of development indicators can vary according to the spatial level of aggregation of the data and the applied methods.

It has been discovered that larger geographic units tend to mask internal heterogeneity of development indicators between areas (Tarkiainen et al., 2010). A central plank of this book is that intelligence gathering at local geographic units is important for accelerating human population development and progress across developing regions. However, it is recognized that it can be difficult to embark upon local-level analysis for generating reliable results where data availability and access are problematic.

Small area geodemographic classifications can serve as a breadth of fresh air where there is a need to investigate subnational population disparities, particularly in data-scarce environments. A classification system that is developed from data with national or near-national coverage for small or local areas can be used as a framework for distinguishing development indicators across small areas according to the typology of areas. By nature, most surveys used to monitor development goals do not have national coverage (WHO, 2018). However, when such surveys are linked to small area classifications that have national coverage, it is possible to generate initial insight into the fundamental characteristics of respondents to survey. Extrapolation and interpolation are techniques used to estimate hypothetical values for a variable based on other observations (Brezinski, 2001). Geodemographic classifications embrace techniques that allow information from thinly based research surveys to be interpolated or extrapolated to local levels (Webber and Longley, 2003; Ojo and Ezepue, 2017). Estimates are generated based on the assumption that people who are more similar in their geodemographic characteristics reside in the same local area, and that those types of local areas will be spread across the country (Vickers and Rees, 2006).

Progress toward meeting many of the global development goals (MDGs and SDGs) is assessed based on survey data collection. However, it is almost impossible to come across country progress reports that present analysis conducted across smaller geographic units. This gives the impression that key development stakeholders may not be particularly interested in grassroots progress toward meeting the targets of these development goals. Adopting small area geodemographic classifications as the basis of generating insights into subnational disparities offers technocrats an alternative way of investigating inequalities in development progress. Furthermore, the use of small area classifications affords those with less technical capabilities the opportunity to rapidly understand and interpret spatial patterns of population disparities within local communities.

3.2 Evidence-Based Decision-Making

There are a number of cultural shifts taking place in the field of development studies for which quantitative skills are critical. First, there is increasing focus on quantitative skills in the training of experts and policymakers (Hersh et al., 2010). Secondly, the increased emphasis on evidence-based practice in global development (Cairney, 2016) requires effective methods for evaluating global development innovations. The development and adaptation of small area classifications uses this impetus to embed quantitative skills into development practice in a way that is immediately relevant to development practitioners and researchers.

Although the phrase "evidence-based decision-making" is broadly used and accepted, it is useful to provide a working definition relevant to the content of this book. According to Davies (2004, p. 3), evidence-based decision-making is about helping

> people make well informed decisions about policies, programs, and projects by putting the best available evidence from research at the heart of policy development and implementation. This approach stands in contrast to opinion-based policy, which relies heavily on either the selective use of evidence (for example, on single studies irrespective of quality) or on the untested views of individuals or groups, often inspired by ideological standpoints, prejudices, or speculative conjecture.

The evidence-based approach is a sharp contrast to making policy decisions based on opinions. Opinion-based policymaking relies on the selective use of evidence. It relies heavily on untested views of individuals or groups of people. Opinion-based decision-making has hindered progress in many developing countries (Mozafarpour, 2011) where approaches to policy decisions have been driven by ideological standpoints, prejudices, or speculative conjecture.

Embracing the use of small area geodemographic classifications can help promote evidence-based practice by using robust empirical evidence to support specific policy interventions in developing countries. For instance, the use of small area classifications might be wrapped around a tripartite framework of:

- Targeting of an intervention (e.g. to an appropriate population, location, or time);
- Testing to understand the efficacy of an intervention; and
- Tracking to measure the problem to be solved and subsequent results of an intervention.

Each of these aspects of evidence-based practice can involve masses of data influenced by spatial and temporal factors. This is an initial challenge to many practitioners working in developing country contexts who are attempting to implement evidence-based practices. Geodemographic classifications have proven value as a useful base for simplifying complex spatial and temporal data for the purposes of determining where resources should be focused, evaluating whether interventions are working, and examining whether practitioners and policymakers are working in accordance with good practice guidelines (Abbas et al., 2009).

There are other ways in which geodemographic classifications can be used to bolster evidence-based decision-making in development policy. Davies (2004) identified a range of evidence types that are commonly generated during the process of policy research. Figure 3.1 illustrates the different types of evidence and how small area classifications might be used to generate them.

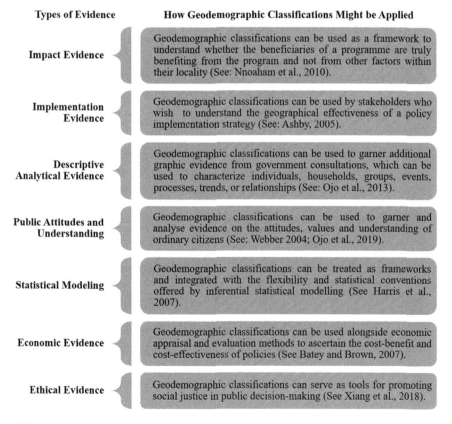

Types of Evidence   **How Geodemographic Classifications Might be Applied**

Impact Evidence — Geodemographic classifications can be used as a framework to understand whether the beneficiaries of a programme are truly benefiting from the program and not from other factors within their locality (See: Nnoaham et al., 2010).

Implementation Evidence — Geodemographic classifications can be used by stakeholders who wish to understand the geographical effectiveness of a policy implementation strategy (See: Ashby, 2005).

Descriptive Analytical Evidence — Geodemographic classifications can be used to garner additional graphic evidence from government consultations, which can be used to characterize individuals, households, groups, events, processes, trends, or relationships (See: Ojo et al., 2013).

Public Attitudes and Understanding — Geodemographic classifications can be used to garner and analyse evidence on the attitudes, values and understanding of ordinary citizens (See: Webber 2004; Ojo et al., 2019).

Statistical Modeling — Geodemographic classifications can be treated as frameworks and integrated with the flexibility and statistical conventions offered by inferential statistical modelling (See Harris et al., 2007).

Economic Evidence — Geodemographic classifications can be used alongside economic appraisal and evaluation methods to ascertain the cost-benefit and cost-effectiveness of policies (See Batey and Brown, 2007).

Ethical Evidence — Geodemographic classifications can serve as tools for promoting social justice in public decision-making (See Xiang et al., 2018).

FIGURE 3.1
Using geodemographic classifications to generate different types of evidence.

3.3 Transparent Resource Allocation

Public administration institutions in developing countries are important democratic podiums for promoting judicious use of national resources. These institutions are often advocated as a means to introduce greater probity and controls on corruption in order to facilitate balanced economic development and to promote public accountability (Kieghe, 2016).

The underlying basis for improved accountability is transparency in the disbursement of public resources. In this book, the interest is in transparent geographic disbursement of resources. There are various reasons as to why transparency is essential for developing countries. First, local, regional, and national public institutions depend on the goodwill of public participation to be effective (Michels and De Graaf, 2010). Second, it is important to acknowledge that a number of developing countries continue to receive financial or material aids from international donor agencies. Those countries that benefit from development cooperation have a moral responsibility to demonstrate transparency and accountability to donor agencies. Third, evidence shows that diminished levels of transparency contribute to a crisis of democracy around the world (Transparency International, 2019).

Figure 3.2 shows a scatterplot of Transparency International's Corruption Perception Index (CPI) versus Fund for Peace's Fragility States Index (FSI). The data used are for 2018. The analysis yields a correlation coefficient of

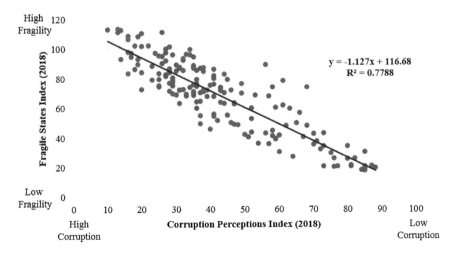

FIGURE 3.2
Scatterplot of the Corruption Perception Index (2018) and the Fragile States Index (2018) (Author's elaboration based on data from Transparency International Corruption Perception Index and the Fund for Peace Fragile States Index).

−0.8, which suggests that state fragility increases as public perception of corruption (used as a proxy for transparency deficit) diminishes.

There are multiple factors that make it challenging for institutions in developing countries to pursue transparency when trying to balance the geographic costs and benefits of their policies and programs. The centralized nature of many governments fosters secrecy and widespread abuse of power in official circles (Edoun, 2012). Furthermore, when trying to decide on the benefits of policies and programs, local public officials often need to contend with multiple conflicting objectives. This is also connected with the fact that public decision-makers are typically presented with a range of opportunities, sometimes with insufficient details about the social and spatial implications. This makes it challenging for some of these institutions to make well-informed decisions.

Geographic equity in resource allocation is one of the main benefits of pursing an ethos of transparency. It requires that communities and individuals with similar needs have access to the same resources. This also needs to be balanced with the scale of the need such that those communities and individuals with greater need have access to proportionally greater resources. For developing countries, judicious allocation of resources is important in many respects and citizens need to be convinced that governments are not shortchanging them. The economic resources required to sustain the teeming population of these countries are also limited and they require fair management and a lot of political will. Although geodemographic classifications cannot be used as a substitute for political will, these can be used to provide decision-makers with much more holistic information about granular population characteristics in a simplified way.

Public sector agencies are often comfortable with making decisions based on univariate analysis. Univariate analysis is the simplest form of data analysis where the data being analyzed contain only one variable. Since it is a single variable, it does not deal with causes or relationships. The main purpose of univariate analysis therefore is to describe the data and find patterns that exist within it. In many instances however, a fusion of multiple indicators has the ability to provide a more holistic picture and reveal something new. Geodemographic approaches allow for the synthesis of multivariate information relevant for better decision-making (Harris et al., 2005). Furthermore, a useful feature of geodemographic classifications is the textual and graphical information that accompany the classifications. These textual and graphical descriptions are often referred to as pen-portraits, and they are used to summarize the predominant attributes of the population groups. They help to elucidate (in qualitative terms) the information generated from complex quantitative analysis. Consequently, when public sector data sets are linked with geodemographic classifications, decision-makers are not only informed about the relationships between different indicators, but also provided with possible pathways toward solutions.

3.4 Targeting of Policy Interventions

Some scholars have contended that the poorest people in developing societies have benefited most from global development aid, and that this has helped to forge progress toward meeting the targets of some global development goals (Minoiu and Reddy, 2009). Others however are more skeptical. They are of the view that a key reason as to why several developing regions of the world have lagged behind others is the long-held assumption that global aid funding and intervention programs reach the poorest segments of society (Browne, 2006). It has been argued that growth and development have remained stunted in certain developing countries because of a misconception that policy interventions are reaching those who need help (Groves and Hinton, 2004). Intelligent allocation of development interventions is essential for reducing or removing inequalities and all the unfairness it creates.

Universalism and targeting are two broad schools of thought used to consider how policies, programs, interventions, and accompanying resources should geographically be allocated to citizens in developing countries. Universalism focuses on the broad spectrum of society and it proposes that all citizens of a country receive the same publicly provided benefits (Anttonen, 2012; Carey et al., 2015). Proponents of universalism often argue that filtering out those communities with the greatest need entails a complex, imprecise, and often inaccurate procedure. Furthermore, some have suggested that universalism is better for developing countries because it reduces additional analytical burdens and associated costs (Giraudy and Pribble, 2019). It is also possible for the systematic targeting of social development interventions in developing countries to generate tensions between communities that benefit and those that do not receive support. This can happen where public officials fail to properly communicate their decision-making rational with local communities.

Despite the merits often highlighted by proponents of the universalist model, this approach suffers from several drawbacks. It fails to recognize that different people and communities will have different needs at different points in time. It also does not account for differences in the scale or size of needs for intervention. Furthermore, universalism struggles where the resources needed for intervention are scarce or minimal. Spreading resources thinly on a universal basis may not achieve the best value for money for citizens (Gugushvili and Hirsch, 2014).

On the contrary, proponents of the geographic targeting of public policy interventions argue for using various mechanisms to identify and distribute the bulk of national resources to the most disadvantaged citizens (Sharp, 2001; Slater and Farrington, 2009). Modern technologies and advancements in empirical techniques are helping to greatly minimize errors when determining who or where to target. The targeted approach is about enhancing equity within communities. It recognizes that some citizens and communities are furthest behind and require more of resources to catch up, succeed, and, eventually, close development gaps.

Geodemographic classifications have been used as tools for enhancing the targeting of workable public interventions (Batey and Brown, 2007). This is important for any government that seeks to reduce cost on the long run. The process of targeting interventions requires mechanisms for identifying special or vulnerable populations groups. The identification of such groups entails unveiling their attributes and locating their distribution across geographic space. Geodemographic classifications offer users the opportunity to understand the probability of occurrence of a phenomenon within a given population group or at a given location. This is of particular importance if public officials aim to deliver timely and relevant interventions (for instance, against crime or public health) and deploy policies smartly. Geodemographic classifications therefore offer stakeholders in developing societies the opportunity to be proactive rather than reactive. All needs are important but not all needs are common to everyone at the same level or at the same time. Geodemographic classifications can be used for needs assessment to support good practice in deploying resources transparently while also assessing the value for money within different contexts.

3.5 Monitoring the Impacts of National Policies

Developing countries are melting pots of various policy programs, thereby attracting the attention of interventions from local and international stakeholders. These different stakeholders are regularly faced with the task of weighing the progress of their policy programs against established objectives, which the programs seek to achieve (Baker, 2000). Consequently, many developing countries are developing mechanisms to help them track the implementation of policy initiatives as well as their impact.

The geographic monitoring of national policy programs involves gathering of evidence to show what progress has been made in the geographic implementation of programs and projects (Cameron et al., 2016). Essentially, there is an emphasis not just on program inputs, activities, and outputs. Occasionally monitoring can also include the geographic outcomes of policy programs. Geographic monitoring cannot be achieved without the regular collection, analysis, and reporting of information to track whether actual results are being achieved as planned.

Demographic shifts across the developing world have necessitated the pursuit of new and complex policy programs, which have generated challenges for monitoring in developing countries. In different ways, geodemographic classifications can be used as tools for ameliorating some of these problems.

Policy intervention programs are increasingly ambitious involving multiple stakeholders. The scale of these programs requires the use of new methods that can identify groups of places that are similar to one another in order to make sense of an almost endlessly complex picture. Geodemographic classifications are developed to recognize that every small geographic area has a unique combination of characteristics.

The amplified focus on sustainability means that policy programs now need to be monitored over longer periods. A consequence of this is that tools and frameworks for tracking development projects need to be suitable for collecting and assessing information far beyond the implementation period of policy programs. Geodemographic classifications are built to subsume the contextual features of people's local environment (Harris et al., 2005), which make them more durable for longer term geographic monitoring of policy programs. Such monitoring approaches allow for the generation of input-feedback mechanisms that enhance the effectiveness of policymaking.

Still within the scope of monitoring the impact of national policies, geodemographic classifications allow users to benchmark population groups and characteristics (Abbas et al., 2009). This means that the performance of a typology of communities can be contrasted to others and relative to a national performance benchmark. Monitoring can be done by comparing the relative performances of the neighborhood types based on the chosen indicators over time. The benefit of doing this can obviously only be realized where time-series data or longitudinal data sets are available.

3.6 Public Sector Social Marketing

A number of developing countries remain trapped in a syndrome of interlocking characteristics of insecurity, ethnoreligious division, violence, and fragility, which undermine sustained progress for social and economic development (Brinkerhoff, 2016). Efforts aimed at tackling these problems mostly focus on removing conditions that trigger institutional breakdown, but struggle to engage with local, everyday experiences of people at the grass roots. Promoting institutional resilience is essential, but it cannot by itself foster a comprehensive reversal of long-established assumptions and practices, which underpin the existing developmental challenges. This can be realized through a decisive shift in public apathy and collective behavioral change.

Public sector promotion of forms of behavioral change often requires innovative mixed-method approaches for the capture and analysis of new forms of social data. The term "social marketing" is an emerging concept, which has emerged within the domain of promoting and sustaining social and behavioral change among citizens. Social marketing has been described

FIGURE 3.3
Social marketing consumer triangle.

as "the systematic application of marketing alongside other concepts and techniques to achieve specific behavioral goals for a social good" (NSM Centre, 2006 p. 4). This implies that social marketing utilizes commercial marketing approaches to change public behavior. It is based on a number of core principles subsumed within the consumer triangle shown in Figure 3.3. Social marketing interventions place the consumers of public at the center and are based on gaining insights about people and places. As can be seen from the figure, when delivering behavioral change interventions, public policy stakeholders grapple with a number of important strategic and operational demands. Policy stakeholders may wish to know how their understanding of what citizens need can help them to enhance the value of interventions they are offering to citizens. Policy developers may also wish to know what might stop members of particular geographic areas from taking up an intervention offer.

In the world of business and commerce, it is important for the desired product or service to reach the appropriate customer in order for a purchase to be made. Just like profit-seeking commercial organizations, public organizations should be interested in ensuring that their product services (e.g. adequate healthcare, education and poverty alleviation programs, and public safety) reach the appropriate target groups via a relevant channel (Rasul and Roger, 2018). The volume, form, and spatial disaggregation of these public sector commodities are vital for the satisfaction of citizens.

Applied geodemographics embraces the idea of introducing strategic and dynamic targeting techniques, which take cognizance of the types and levels of variations in the needs of the target population (Harris et al., 2005). In the UK, geodemographic classifications have been used extensively to support social marketing particularly in the health sector (Powell et al., 2007). Geodemographic classifications enable service providers to know who their customers are, what they do, and what their general attitudes are (Leventhal, 2016). It can help to uncover segments of the population that had been difficult to reach in the past. The classifications also have the benefits of allowing service providers to vary their communication channels and set differing objectives based on the dynamics of the target population groups.

Another social marketing strength of geodemographic classifications is that they help to promote policies based on a multidimensional understanding of people's lives (Williams and Botterill, 2006). Doing this can help make the marketing and ultimate impact of social development policy goals realistic and achievable. Policymakers are also keen to ensure that they are smarter in terms of their overall promotional strategy especially where resources are scarce. Geodemographic classifications can be used to help stakeholders target their social marketing resources cost-effectively and select those interventions that have the best impact over time (Farr et al., 2008). By segmenting people and places into groups based on multicriteria analysis, geodemographic classifications allow public sector service providers to develop products, services, and communications that fit people's needs and motivations.

3.7 Differential Communication Strategies

One of the most important activities in the public policy implementation cycle across developing countries is the need to determine the most appropriate channel for communicating behavioral change or interventions to different groups of people (Snyder, 2007). A choice of the appropriate method of communicating with people can be very daunting especially in developing countries where the uptake of public services is often associated with trust in public service providers. The method of deciding on the communication channel to employ for the delivery of public policy in developing countries can be greatly enhanced by taking a holistic appraisal of the differential characteristics of local populations (Gurman et al., 2012).

Small area geodemographic classifications are generally built on a comprehensive understanding of citizens as consumers of public services, and the drivers of behavioral change at individual and societal levels (Williams

and Botterill, 2006). Many modern geodemographic classifications subsume measures of multiple communication channels. These channels of
communication include traditional forms of public education channels,
such as mass media advertising, anti-behavior television advertisements,
and radio jingles. However, information about other nontraditional communication channels, such as new media, is also being incorporated into
geodemographic classifications (Chappell et al., 2017) to help enhance
smarter communication.

Another strength of small area geodemographic classifications is that
they subsume citizen's values, which also help with systematic channeling
of communication (Harris et al., 2005). Consumer values and beliefs are
things that consumers hold special to them. Such issues often form their
opinions, influence their decisions, and eventually motivate their actions.
For instance, the perception and response of people to technological growth
and change can inform how technology is used to channel public policy
information to them (Longley et al., 2006). Other prominent indicators that
have been supplanted within geodemographic classifications include the
way people spend their leisure times (indoors and outdoors) (Leventhal,
2016), which can inform the design and communication of public health
interventions, for instance.

Consumption patterns are also included in segmenting people and
places into geodemographic typologies (Dalton and Thatcher, 2015). This
information gives an insight into where people would prefer to stake their
money and why. Some people, for instance, prefer categories of established
products, while others might choose to try out new brands. Consumption
patterns are also indicative of the readership interests (magazines,
newspapers, and telegraphs) of the public and how much audience they
give to the screen.

Indicators of citizen's psychology are also used to generate
geodemographic classifications (Harris et al., 2005). These characteristics
explain the way people relate to the products and services they purchase
or use. Family structure and patterns as well as the different career types
(Mitchell and McGoldrick, 1994) can influence the psychology of people
within small geographic spaces. Questions about the main goals and
desires of people in life are major issues, which affect their psychology,
their behaviors, and ultimately how receptive they are to public policy
communications.

The response and reaction of members of the public to issues relating to
policy marketing can be inferred by understanding their sociology, psychology, and environment. It is difficult to measure these values. However,
with the deployment of geodemographic classifications, the level of citizen
uptake of information via these different channels of communication can be
ascertained for different geographic areas. Such information helps public
policy practitioners to vary their channel of suitable communication on a
geographic basis.

3.8 Geographic Forecasting of Social and Economic Futures

Geographic forecasting is a procedure for generating information about future conditions of geographic areas based on prior information about policy problems (Ballas et al., 2007). Forecasting can be based on the extrapolation of current and historical trends into the future, the use of explicit geographic theoretical assumptions, or expert judgments about future states of society (George et al., 2004).

The role of forecasts in geographic policy decision-making in developing countries cannot be overemphasized. The socioeconomic repercussions of a policy decision made today only unfold over time and the new policy is typically intended to remain in place for some period. Forecasts provide information about future changes in policies and their potential consequences. The ability to forecast permits greater control through understanding past policies and their consequences, which implies that the future is determined by the past. Furthermore, forecasting enables developing countries to shape the future in an active manner, irrespective of what has happened in the past.

To improve the accuracy of geographic forecasts, development policy stakeholders are constantly in search of demand planning frameworks and tools that display inferential capabilities (Nazmi, 2015). Geodemographic classifications are built from methods that use causal variables to define the segments. Consequently, these classifications make it possible to use various sources of a priori information to forecast public behavior within each geodemographic segment. Geodemographic classifications prove advantageous when the effect of one variable is examined before considering the effect of another. This is described as causal priorities. Classifications offer a simple alternative of splitting the data first on the variables that come earliest in the causal chain.

When developing small area geodemographic classifications, different sources of information can be consulted and fused together (Mateos and Agular, 2013). This approach to segmenting geographic areas provides a convenient mechanism for incorporating different dimensions of expert judgment into the classification of areas. Expert judgment has been described as beneficial for forecasting social and economic futures in developing countries particularly when expected values and corresponding prediction intervals are difficult to access based on formal methods (Kielman, 2019).

Another important advantage of those geodemographic classifications that incorporate fuzzy logic is that they do not use generalized assumptions about human behavior and population patterns (See and Openshaw, 2001). Unlike some popular econometric techniques that may assume all people in a geographic region, for instance, interact with a causal indicator in the same manner, small area geodemographic classifications do not generalize assumptions within large geographic regions. The advantage of this is that the detrimental impacts of data-related problems such as interaction, nonlinear effects, and causal priorities can be reduced when forecasting. For instance, a policy intervention aimed at reducing domestic violence may be appropriate for adults in a particular geographic jurisdiction, but it may prove counterproductive for adults in another jurisdiction. Rather than focus on the effect of the intervention alone, geodemographic classifications would identify groups that respond in given ways to this policy mediation.

3.9 Conclusion

While the deployment of small area geodemographic classifications in the public sector has lagged behind commercial applications in the developed world, there is evidence that public sector use of these classifications is growing. However, it is very rare to come across the augmentation of public policy with small area classifications in developing countries. The adaptation of the discriminatory power of geodemographic classifications within the public sector often sparks interests in issues related to resource targeting. The profiles derived from geodemographic typologies are increasingly playing a central role in ensuring effective deployment of resources by public services. Some of the key sectors where area classification typologies have played important roles include public health, community safety, education, emergency planning and response, business and commerce, and politics. Geodemographic classifications have enhanced decision-making within these sectors because of their capacity to provide simplified representations of the mass of detailed counts of data held in small area statistics. As demonstrated in this chapter, developing countries will find small area classifications useful by linking these tools with achieving local and global policy agendas such as the SDGs. These classifications can help to increase the understanding of underlying trends and policy problems, including what is driving area changes and diverging area fortunes, and whether the characteristics of small geographic areas themselves make a difference in individual outcomes.

References

Abbas, J., Ojo, A., Orange, S. (2009). Geodemographics – A Tool for Health Intelligence. *Public Health*, 123(1), 35–39.

Anttonen A. (2012). Universalism and the Challenge of Diversity. In: A. Anttonen, L. Haikio and K. Stefansson (Eds), *Welfare State, Universalism and Diversity*. Cheltenham: Edward Elgar.

Ashby, D.A. (2005). Policing Neighborhoods: Exploring the Geographies of Crime, Policing and Performance Assessment. *Policing and Society*, 15(4), 413–447.

Batey, P. and Brown, P. (2007). The Spatial Targeting of Urban Policy Initiatives: A Geodemographic Assessment Tool. *Environment and Planning A*, 39(11), 2774–2793.

Baker, J.L. (Ed). (2000). *Evaluating the Impact of Development Projects on Poverty: A Handbook for Practitioners*. Washington, DC: World Bank.

Ballas, D., Clarke, G., Dorling, D. and Rossiter, D. (2007). Using SimBritain to Model the Geographical Impact of National Government Policies. *Geographical Analysis*, 39(1), 44–77.

Brezinski, C. (2001). *Interpolation and Extrapolation, Volume 2 (Numerical Analysis 2000)*. Amsterdam: North Holland.

Brinkerhoff, D.W. (2016). State Fragility, International Development Policy, and Global Responses. In: D. Stone and K. Moloney (Eds), *The Oxford Handbook of Global Policy and Transnational Administration*. Oxford: Oxford University Press.

Browne, S. (2006). *Aid and Influence: Do Donors Help or Hinder?* Abingdon: Earthscan.

Cairney, P. (2016). *The Politics of Evidence-Based Policy Making*. London: Palgrave Macmillan.

Cameron, D.B., Mishra, A. and Brown, A.N. (2016). The Growth of Impact Evaluation for International Development: How Much Have We Learned? *Journal of Development Effectiveness*, 8(1), 1–21.

Carey, G., Crammond, B. and De Leeuw, E. (2015). Towards Health Equity: A Framework for the Application of Proportionate Universalism. *International Journal for Equity in Health*, 14(81), 1–8.

Chappell, P., Tse, M., Zhang, M. and Moore, S. (2017). Using GPS Geo-Tagged Social Media Data and Geodemographics to Investigate Social Differences: A Twitter Pilot Study. *Sociological Research Online*, 22(3), 38–56.

Dalton, C.M. and Thatcher, J. (2015). Inflated Granularity: Spatial "Big Data" and Geodemographics. *Big Data and Society*, 2(2): 1–15.

Davies, P. (2004). *Is Evidence-based Government Possible?* Jerry Lee Lecture presented at the fourth Annual Campbell Collaboration Colloquium, Washington DC.

Edoun, E.I. (2012). Decentralization and Local Economic Development: Effective Tools for Africa's Renewal. *International Journal of African Renaissance Studies – Multi-, Inter- and Transdisciplinarity*, 7(1), 94–108.

Farr, M., Wardlaw, J. and Jones, C. (2008). Tackling Health Inequalities Using Geodemographics: A Social Marketing Approach. *International Journal of Market Research*, 50(4), 449–467.

George, M.V., Smith, S.K., Swanson, D.A. and Tayman, J. (2004). Population Projections. In: J. Siegel and D. Swanson (Eds), *The Methods and Materials of Demography*. San Diego, CA: Elsevier Academic Press.

Giraudy, A. and Pribble, J. (2019). Rethinking Measures of Democracy and Welfare State Universalism: Lessons from Subnational Research. *Regional and Federal Studies*, 29(2), 135–163.

Gurman, T.A., Rubin, S.E. and Roess, A.A. (2012). Effectiveness of mHealth Behavior Change Communication Interventions in Developing Countries: A Systematic Review of the Literature. *Journal of Health Communication*, 17(1), 82–104.

Groves, L. and Hinton, R. (Eds). (2004). *Inclusive Aid: Changing Power and Relationships in International Development*. Abingdon: Earthscan.

Gugushvili, D. and Hirsch, D. (2014). *Means-Testing or Universalism: What Strategies Best Address Poverty?* Loughborough: Centre for Research in Social Policy.

Harris, R., Johnston, R. and Burgess, S. (2007). Neighborhoods, Ethnicity and School Choice: Developing a Statistical Framework for Geodemographic Analysis. *Population Research and Policy Review*, 26(5), 553–579.

Harris, R., Sleight, P. and Webber, R. (2005). *Geodemographics, GIS and Neighborhood Targeting*. London: Wiley.

Hersh, W., Margolis, A., Quirós, F. and Otero, P. (2010). Building a Health Informatics Workforce in Developing Countries. *Health Affairs*, 29(2), 274–277.

Jacob, A. (2017). Mind the Gap: Analyzing the Impact of Data Gap in Millennium Development Goals' (MDGs) Indicators on the Progress toward MDGs. *World Development*, 93, 260–278.

Kieghe, D. (2016). *National Ambition: Reconstructing Nigeria*. London: New Generation Publishing.

Keilman N. (2019). Erroneous Population Forecasts. In: T. Bengtsson and N. Keilman (Eds), *Old and New Perspectives on Mortality Forecasting. Demographic Research Monographs (A Series of the Max Planck Institute for Demographic Research)*. Cham: Springer.

Leventhal, B. (2016). *Geodemographics for Marketers: Using Location Analysis for Research and Marketing*. London: Kogan Page.

Liverman, D.M. (2018). Geographic Perspectives on Development Goals: Constructive Engagements and Critical Perspectives on the MDGs and the SDGs. *Dialogues in Human Geography*, 8(2), 168–185.

Longley, P., Ashby, D., Webber, R. and Li, C. (2006). Geodemographic Classifications, the Digital Divide and Understanding Customer Take-Up of New Technologies. *BT Technology Journal*, 24(3), 67–74.

Macfarlane, S.B., AbouZahr, C. and Tangcharoensathien, V. (2019). National Systems for Generating and Managing Data for Health. In: S. Macfarlane and C. AbouZahr (Eds), *The Palgrave Handbook of Global Health Data Methods for Policy and Practice*. London: Palgrave Macmillan.

Mateos, P. and Aguilar, A.G. (2013). Socioeconomic Segregation in Latin American Cities: A Geodemographic Application in Mexico City. *Journal of Settlements and Spatial Planning*, 4(1), 11–25.

Michels, A. and De Graaf, L. (2010). Examining Citizen Participation: Local Participatory Policy Making and Democracy. *Local Government Studies*, 36(4), 477–491.

Minoiu, C. and Reddy, S.G. (2009). Development Aid and Economic Growth: A Positive Long-Run Relation. *The Quarterly Review of Economics and Finance*, 50(1), 27–39.

Mitchell, V. and McGoldrick, P. (1994). The Role of Geodemographics in Segmenting and Targeting Consumer Markets. *European Journal of Marketing*, 28(5), 54–72.

Mozafarpour, S., Sadeghizadeh, A., Kabiri, P., Taheri, H., Attaei, M. and Khalighinezhad, N. (2011). Evidence-Based Medical Practice in Developing Countries: The Case Study of Iran. *Journal of Evaluation in Clinical Practice*, 17, 651–656.

Nazmi, N. (2015). *Economic Policy and Stabilization in Latin America*. Abingdon: Routledge.

Nnoaham, K.E., Frater, A., Roderick, P., Moon, G. and Halloran, S. (2010). Do Geodemographic Typologies Explain Variations in Uptake in Colorectal Cancer

Screening? An Assessment Using Routine Screening Data in the South of England. *Journal of Public Health*, 32(4), 572–581.

NSM Centre (2006). *Social Marketing Works: A Short Introduction for NHS Staff*. London: National Social Marketing Center.

Ojo, A. and Ezepue, P.O. (2011). How Developing Countries can Derive Value from the Principles and Practice of Geodemographics, and Provide Fresh Solutions to Millennium Development Challenges. *Journal of Geography and Regional Planning*, 4(9), 505–512.

Ojo, A., and Ezepue, P.O. (2017). Using Casualty Assessment and Weighted Hit Rates to Calibrate Spatial Patterns of Boko Haram Insurgency for Emergency Response Preparedness. *Journal of Terrorism Research*, 8(4), 1–7.

Ojo, A., Ibeh, S.C. and Kieghe, D. (2019). How Nigeria's 2015 Presidential Election Outcome Was Forecasted with Geodemographics and Public Sentiment Analytics. *African Geographical Review*, 38(4), 343–360.

Ojo, A., Vickers, D., Ballas, D. (2013). Creating a Small Scale Area Classification for Understanding the Economic, Social and Housing Characteristics of Small Geographical Areas in the Philippines. *Regional Science Policy and Practice*, 5(1), 1–24.

Powell, J., Tapp, A., Sparks, E. (2007). Social Marketing in Action – Geodemographics, Alcoholic Liver Disease and Heavy Episodic Drinking in Great Britain. *International Journal of Nonprofit and Voluntary Sector Marketing*, 12(3), 177–187.

Rasul, I. and Rogger, D. (2018). Management of Bureaucrats and Public Service Delivery: Evidence from the Nigerian Civil Service. *The Economic Journal*, 128(608), 413–446.

Sharp, K. (2001). *An Overview of Targeting Approaches for Food-Assisted Programming*. Atlanta, GA: CARE USA.

See, L. and Openshaw S. (2001). Fuzzy Geodemographic Targeting. In: G. Clarke and M. Madden (Eds), *Regional Science in Business. Advances in Spatial Science*. Berlin: Springer.

Slater, R. and Farrington, J. (2009). *Targeting of Social Transfers: A Review for DFID*. London: Overseas Development Institute.

Snyder, L.B. (2007). Meta-Analyses of Mediated Health Campaigns. In: R.W. Preiss, B.M. Gayle, N. Burrell, M. Allen and J. Bryant (Eds), *Mass Media Effects Research: Advances through Meta-Analysis*. Mahwah, NJ: Lawrence Erlbaum Associates Publishers.

Tarkiainen, L., Martikainen, P., Laaksonen, M. and Leyland, A.H. (2010). Comparing the Effects of Neighborhood Characteristics on All-Cause Mortality Using Two Hierarchical Areal Units in the Capital Region of Helsinki. *Health and Place*, 16(2), 409–412.

Transparency International (2019). *Corruption Perceptions Index 2018*. Berlin: Transparency International.

Vickers, D. and Rees, P. (2006). Introducing the Area Classification of Output Areas. *Population Trends*, 125, 15–24.

Webber, R. (2004). *The Relative Power of Geodemographics Vis a Vis Person and Household Level Demographic Variables as Discriminators of Consumer Behavior*. London: Centre for Advanced Spatial Analysis.

Webber, R. and Longley, P.A. (2003). Geodemographic Analysis of Similarity and Proximity: Their Roles in the Understanding of the Geography of Need.

In: P.A. Longley and M. Batty (Eds), *Advanced Spatial Analysis*. Redlands, CA: Environmental Systems Research Institute Inc. (ESRI) Press.

WHO (2018). *Summary Report on the Expert Consultative Meeting to Discuss Priority National Population-Based Surveys for Better Reporting of WHO Regional Core Indicators and SDG Health-Related Indicators*. Cairo: World Health Organization. Regional Office for the Eastern Mediterranean.

Williams, S. and Botterill, A. (2006). Profiling Areas Using the Output Area Classification. *Regional Trends*, 39, 11–18.

Xiang, L., Stillwell, J., Burns, L., Heppenstall, A. and Norman P. (2018). A Geodemographic Classification of Sub-Districts to Identify Education Inequality in Central Beijing. *Computers, Environment and Urban Systems*, 70, 59–70.

4

Reasons for Slow Proliferation of Area Classification across Developing Countries

4.1 Lack of Publicity of the Benefits of Area Classifications

A detailed discussion of the benefits and potential of small area geodemographic classifications has been provided in the third chapter of this book. A careful examination of the discussion shows that contemporary interest in the development and deployment of geodemographic problem-solving approaches has evolved and exploded across many parts of the developed world (Willis et al., 2010; Kimura et al., 2011; Singleton and Spielman, 2014). Various sectors and industries have found geodemographic classifications beneficial for problem definition, structuring, and solving. These include health (Farr and Evans, 2005; Shelton et al., 2006; Abbas et al., 2009), education (Singleton, 2010; Singleton et al., 2012), policing (Ashby and Longley, 2005), local and public government (Brown et al., 1999); energy (Goodwin and Sykes, 2012), human population mobility (Sabater, 2015), electoral dynamics (Webber, 2005), and commercial and financial services (Leventhal, 2016).

Despite a growing body of academic and nonacademic materials on geodemographics emerging in the northern hemisphere, the use and adaptation of these area classifications is very rare in developing countries. Since the underlying techniques for developing these area classifications emerged within Anglo-American settings (Harris et al., 2005), it is natural that countries in the northern hemisphere benefited first from the growth and explosion of geodemographic ideas. However, the benefits of these area classifications for developing countries have never really been widely publicized across the southern hemisphere. This has contributed toward the slow proliferation of geodemographic classifications across the region. Until the publication of this book, no real effort has been made to juxtapose the potential of small area classifications to a careful diagnostic of the main challenges and opportunities exposed to citizens and policymakers in developing regions.

In a sense, countries in developing regions seem to have been accidentally alienated from the refreshing debates and development of area classifications. One reason for this is that original applications of area classifications focused on the business and commercial sector (Leventhal, 2016). The competition between commercial organizations meant that they embraced a black box approach. The majority of openly available public policy applications in the northern hemisphere began to emerge toward the middle of the first decade of the present millennium.

Given that the key priorities confronting the vast majority of developing countries are humanitarian and development-centered issues (ODI, 2017), it has been difficult to promote the benefits of area classifications in the developing world using existing commercial or profit-centric examples based on black box approaches from the northern hemisphere. However, now that public sector deployment of geodemographic classifications is gaining ground in developed countries, it becomes imperative to highlight how these techniques can also be deployed for human-centered problem-solving and policymaking in developing countries.

The internationalization of social research practice is increasingly desirable (Truong, 2015). There is broad recognition that the challenges and opportunities confronting an increasingly globalized world stand to benefit from the exchange of ideas from multiple contexts. Anglo-American dominance of the geodemographic research practice space does not allow for the cross-fertilization of critical ideas from other national and linguistic borders. For instance, how do we know that these techniques really work in other contexts if scholars within those jurisdictions are not brought into the debate? As long as developing countries remain alienated from the adaptation of small area geodemographic classifications, this type of research and policy practice cannot be said to be truly international.

The interdisciplinary nature of the application of geodemographic classifications (Webber and Burrows, 2018) means that any publicity agenda should seek to make research findings more broadly accessible. Although academic journals tend to be the traditional media for publicizing emerging and evolving research practices, the publicity of the benefits of small area classifications across developing regions must extend beyond publishing materials in peer-reviewed journals alone. This is necessary because a substantial part of the audience that may embrace and benefit from these classifications include practitioner and nonacademic communities.

Greater publicity of the benefits of small area classifications across the developing world will attract good graduate students from these regions. Enhanced publicity will also stimulate informal networking that often leads to opportunities for formal collaborations. Dedicated publicity is also likely to get the interest from the private sector, policymakers and nongovernmental organizations (NGOs) that are working at the vanguard of human development.

4.2 Institutional and Organizational Weaknesses

Small area classifications have been designated as innovative solutions for improving the efficacy of targeted human-centered interventions (Brown et al., 1999). It is rare for countries to succeed in the absence of agile state institutions. Therefore, a successful innovation diffusion often depends primarily on the existence of institutions and organizations that have the capability to establish rules, enforce them, generate revenues, and provide services to the public (Chang, 2011). However, policymakers in several developing countries still grapple with how they can acquire high-quality public sector institutions (Brinkerhoff and Brinkerhoff, 2015).

Statistical agencies and institutions in developing countries are integral organizations that can help to accelerate the proliferation of small area classifications within these regions. However, a vast number of these agencies are constrained by weaknesses of institutional and organizational setup. They also often lack the resources and infrastructure needed to accelerate innovation diffusion. Consequently, their performances are often not up to mark.

Evidence from diverse sources shows that some public sector statistical agencies suffer serious deficiencies in establishing law and order within their organizations (Jerven, 2013). Although formal rules and procedures are created and communicated through various channels within these agencies, these rules are rarely enforced. A lack of enforcement of institutional rules within these statistical agencies fosters dysfunctional environments. This leads to the emergence of unwritten socially shared rules, which are systematically enforced outside officially sanctioned channels. Chaotic statistical institutional environments with deficiencies in the establishment of law and order often frustrate attempts to introduce new ideas and new ways of thinking.

4.3 Cultural Constraints

Another factor that makes formal institutions less receptive to new ideas, such as small area classifications, is their philosophical posturing based on cultural constraints. It has been argued that there is an intrinsic link between the culture of organizations and the prevailing national or societal culture (Hofstede and Peterson, 2000). In a vast number of developing societies, hierarchy implies an existential inequality (Wilkinson, 1997; Buchmann and Hannum, 2001). This creates an atmosphere where subordinates expect to be told what to do and the ideal boss is a benevolent autocrat (Kieghe, 2016).

This propagates a culture of cronyism within organizations, which stifles meritorious diffusion of ideas.

Collectivistic cultures emphasize the needs and goals of the group as a whole over the needs and desires of each individual (Han et al., 2015). In such cultures, relationships with other members of the group and the inter-connectedness between people play a central role in each person's identity. Some researchers have argued that cultures in Asia, Central America, South America, and Africa tend to be more collectivistic (Triandis, 2015). This presents a number of implications within and across organizations. First, it means that different value standards are applied to people within and out-side groups because of greater emphasis on common goals. Although it has been suggested that collectivist cultures may generate equal opportunities for all (Fadil et al., 2005), collectivism, according to some, also leads to dis-crimination of opportunities. Collectivist organizational cultures make it difficult to build relationships with new people. In these organizations, rela-tionship prevails over task.

Another factor acting as a cultural constraint against the proliferation of small area classifications in developing countries is that some coun-tries in these regions are perceived as masculine societies where different expectations are stressed for males and females (Pasura and Christou, 2018). Consequently, it is generally anticipated that within organizations, males will be more assertive and competitive. Unlike those organizations that have embraced cultures where gender roles are more fluid, masculine cultures restrict the prospects of females. This does not allow organizations to reach their full potential. People within such organizations may be less receptive to new ideas, such as embracing the use of small area classifications, if a female presents such an idea.

4.4 Physical Infrastructure versus Spatial Data Infrastructure

Physical infrastructure encompasses the basic facilities, systems, and structures required for an economy to function and survive, such as trans-portation networks, a power grid, and sewerage and waste disposal systems (Jimenez, 1995). It is widely recognized that inadequate provision of physical infrastructure has lowered the quality of life of citizens in developing regions and held back the pace of development (Du Plessis, 2007). Fast-growing cities in some African, Asian and Latin American countries are currently afflicted by regular traffic jams due to poor road conditions, power cuts, and housing and water shortages (Roberts, 2005; Vairavamoorth, 2008; Sietchiping, 2012).

In developed countries, there is already a legacy of physical infrastructure investment. Furthermore, the populations in these economies are growing at a much slower pace. Consequently, the priority of northern hemisphere economies focuses on the maintenance and improvement of existing infrastructure as well as acceleration toward green infrastructure (OECD, 2015). It is a different story across developing countries where there has been a deficit in infrastructure investment. Several developing economies are therefore starting from scratch. At the same time, the populations of these regions are growing at an unprecedented rate, outpacing the rate of investment in infrastructure in some cases (Eberhard and Shkaratan, 2012). As a result of these competing demands, the provision of physical infrastructure across developing countries is almost by definition the basis for development.

The emphasis placed on the provision of physical infrastructure in developing countries has proven somewhat counterproductive, as minimal attention seems to have been paid to other pillars of growth and development, such as spatial data infrastructure (SDI) (Makanga and Smit, 2010). Therefore, most sectors of developing economies now increasingly rely on SDI; it should be treated as a public infrastructure that enables the creation of a wide range of products and services. Just as developing societies strategically plan, invest in, and maintain the physical infrastructure they rely on, it is also imperative to begin to do the same with SDI. However, investment in SDI in some developing countries is still very much below par. The shortage of investment means that these data infrastructures lack some basic capabilities. For instance, the enablement of online access to a wide range of spatial information and services in a host of developing countries is still a problem (Indrajit et al., 2019). Similarly, Ojo and Ezepue (2012) have identified that there are weaknesses in enabling integration of geographically distributed spatial information.

The development of small area classifications relies on the ability to synchronize data. However, it has been argued that existing SDIs in some developing countries do not effectively enable collaboration by multilateral information exchange and synchronization (Schwarte, 2008). Furthermore, existing data infrastructures need to become more agile, thereby allowing autonomous organizations to develop interdependent relationships in a distributed environment. This may help to facilitate the definition and sharing of spatial semantics.

The stated capability weaknesses are further exacerbated by the problem of interoperability. Interoperability facilitates information sharing and allows users to find information, services, and applications when needed, independent of physical location (Sondheim et al., 1999). It allows users to understand and employ the discovered information and tools regardless of whether the platform is local or remote. Furthermore, through interoperability, users can also evolve a processing environment without being constrained to a single vendor's offerings (Vckovaki, 1998).

Despite well-documented progress made in bridging the digital divide between developed and developing countries (Chiemeke, 2010), there remain significant problems in terms of conformance and interoperability due to the dearth of human capacity and of training opportunities, standardization, testing, and certification.

SDI interoperability in some developing countries is also compounded by the fact that ICT equipment in some public parastatals are old. These equipment are yet to be withdrawn from use due to limited capital to purchase replacements (Weerakkody et al., 2007; Qaisar et al., 2010). The connection interfaces and protocols of such older systems are not able to communicate with some modern systems that are more complex and sophisticated. Gateways are used to facilitate connection and this reduces functionality and increases costs.

Having agile SDIs in developing countries will become more vital as their populations grow and their economies and societies become ever more reliant on getting value from data to meet a range of development needs.

4.5 Analytical Capacity and Costs

Dearth in analytical capacity is another constraint on the proliferation of small area classifications. Although few statistical agencies in developing countries like to admit this problem, many of them are undermined by technical skills gaps in areas of creating, storing, manipulating, and management of spatial statistics (Tumba and Ahmad, 2014). Support is often received from international donor agencies.

The underlying techniques for developing and deploying small area classifications as presented in Chapters 5, 6, and 7 of this book imply a wide range of analytical demands. Although great effort has been made to simplify the sophisticated models and techniques in order to make them user-friendly, it still requires intermediate to advanced level of spatial statistical understanding to implement and interpret these models and techniques.

NGOs are important development stakeholders in developing countries. They face stiffer analytical problems than some of the statutory development agencies. These NGOs are smaller-scale practitioners, which mostly rely on qualitative work often lacking basic training in statistics. Consequently, they tend to exclude the deployment of quantitative approaches from their work. The development and use of small area classifications is heavily skewed in favor of quantitative approaches, which means that most NGOs that generally embrace qualitative work may struggle with adopting geodemographic techniques.

It is impossible to discuss the dearth in analytical capacity without mentioning the repercussions of migration and brain drain on the analytical landscape of developing countries. The departure of highly skilled technical personnel decimates the human and fiscal revenues of the supplying developing countries (Benedict and Ukpere, 2012). Increased decimation of human capital in developing countries because of highly skilled migration has led to calls for policies that restrict the flow of skilled workers (Kuehn, 2007). There are, however, those who present a contrasting view of the brain drain debate. They argue that a highly educated diaspora can become a potent force for analytical capacity development through knowledge transfers and giveback programs and projects (Meyer et al., 1997). However, the impact of these types of knowledge transfer programs can be frustrated by external factors, such as the provision of an enabling environment, the influence of local politics, and the availability of technical infrastructure.

Also closely associated with the challenges of analytical capacity is cost. As mentioned earlier, the adoption of small area classifications requires analysis that is more sophisticated with an equal measure of data needs. In general, the more sophisticated the analysis, and the more the data requirements, the more expensive it becomes to undertake the entire exercise. Costs typically accumulate from the time spent on understanding, analyzing, and interpreting data. Similarly, high costs are associated with the software packages used for data analysis and visualization. However, the gradual emergence of open-source software packages will help to lower software-related costs in the future.

4.6 Local Security and Safety Conditions

Although recent evidence indicates that the world as a whole is becoming less peaceful (IEP, 2018), it has also been established that most fragile regions on the planet today are located in developing countries (OECD, 2018). Some progress has been made in recent years to improve the understanding of what drives tensions, conflict, and violence in fragile states. This is vital since almost half the world's poor are expected to live in countries affected by fragility, conflict, and violence by 2030 (World Bank, 2016). Nevertheless, fragility continues to present a serious impediment to social and economic development in developing countries.

The Department for International Development (DfID) and the Organisation for Economic Co-operation and Development (OECD) are two of the leading contributors to debates, policy, and programming on this subject matter. According to the DfID, fragility results from a situation where a government is unable or unwilling to deliver core functions to the majority of its population, including the poor (DFID, 2005). Such core functions could include service entitlements, justice, and security. The OECD definition is similar. However, it emphasizes the "lack of political commitment and insufficient capacity to develop and implement pro-poor policies" (Prest et al., 2005, p. 5). It would appear though that these definitions are restricted to conflict or immediate post-conflict countries. However, there are developing countries that are experiencing conflicts but are nonetheless providing an acceptable level of service entitlements to the majority of their population (Akhmouch, 2012).

Conflict, insecurity, and fragility inflict major costs on citizens and countries in terms of loss of lives, human suffering, and foregone development opportunities. In addition to these human costs, conflict makes it difficult to gather necessary data in conflict-afflicted regions of the world (Haer and Becher, 2012). The difficulty of pulling together data in a timely and accessible way under fragile conditions means that decision-making is usually based on anecdotal evidence. Quality data are important for the development of reliable small area classifications, but they are much harder to garner in situations of violence. This contributes toward the slow expansion of small area classifications in conflict-affected regions, as the reliance on anecdotal evidence is insufficient for the deployment of reliable classifications.

4.7 Existence of Data

Policymakers in developing countries are increasingly being urged to embrace evidence-based research to inform development decisions. The evidence-based mantra relies on rigorous collection of data. Data, and especially good-quality data, are important for institutions to accurately plan, fund, and evaluate development initiatives.

Small area classification methods rely on the existence of data that are derived from capital-intensive consultations such as national surveys and human population censuses. These data collection activities often require strong justifications in order to attract capital investment. Almost two decades ago, Dackam (2003) argued as follows:

> Only a population census can give accurate and reliable data and information on the population for each geographic level of the African countries. Alternative methods are not feasible and other sources are not available in sub-Saharan Africa, and cannot be anticipated in the near future. The census is the unique source of information for data on specific populations and sub-populations poorly represented in household surveys.
>
> (Dackam, 2003, p. 96)

Parts of Dackam's (2003) claim remain true today. The census remains the most comprehensive source of population data. However, it seems countries in Africa are conducting fewer censuses as inter-census periods are widening (UNFPA, 2002). The cost of conducting censuses has been rising, which means that recent rounds of millennial censuses have been marked by funding crises (UNFPA, 2002). This problem has been exacerbated by shrinking public sector budgets, which means that certain statutory responsibilities like the conduct of national censuses have either been delayed or cancelled all together.

Lack of contemporaneous census and household survey data constitutes a major problem for the implementation of small-area analytical methodologies. To ameliorate this type of problem, researchers have recommended the combination of multiple data sets (Harris et al., 2005). However, this approach also presents additional problems of coping with differences in the timing and compatibility between databases.

4.8 Access to Data

Securing access to existing data constitutes a barrier for the development of small area classifications in developing countries. In some countries, relevant data sets are systematically constricted by the same sources from which they should be derived (Ojo and Ezepue, 2011).

Another factor that makes it difficult to access small area statistics is that some countries still dwell on outdated legal frameworks for the dissemination of information (Eurostat, 2013). In some countries, the deployment of digital or electronic data dissemination channels remains at an infancy stage. Often public sector agencies hide under the cover of personal information disclosure controls rather than making it clear that they lack the digital resources and possibly the manpower to process public information requests.

The problem of accessing data in developing countries can also be linked to certain aspects of societal and sociopolitical influences. Many countries in these regions have grappled with authoritarian and repressive regimes for long periods. Those countries that have transitioned into democratic rule are still emerging from the shadows of authoritarianism, which means that some public agencies are gradually becoming familiar with the idea of freedom of access to information. The parliaments of some countries that were under military rule for long periods have either delayed the passage of freedom of information bills or watered such bills down (Ojo, 2010). This continued secrecy greatly hampers the ease with which researchers can gain access to required data for unbiased analysis. It is important for public office holders to recognize that publicly sourced information like censuses and surveys constitutes a public good. Additionally, making information available for the public signifies that the government is transparent and ready to be accountable and responsible for its actions and policy decisions. Better access to data will allow for constructive debates and discussions among academics, practitioners, and decision-makers.

4.9 Quality of Data

The reliability of results generated from any form of quantitative analysis depends largely on the quality of data upon which the analysis is based. Quality data need to be fit for purpose and should represent the requirements of their authors, users, and administrators. Human population data sets are used for the analysis of present and future demographic patterns and processes irrespective of the geographic scale at which such an analysis is conducted. Quality data are therefore essential not only for making immediate decisions, but also for the purpose of monitoring the performance of interventions and policies, and preempting the future. Data collected for the creation of small area classifications must be accurate, complete, reliable, legible, and accessible to authorized users. However, access to good-quality demographic and spatial data in developing countries has been a persisting challenge for researchers working in these regions for quite some time (WHO, 2003). In order to understand how poor data quality undermines the prospects of small area classifications in developing countries, it is necessary to understand the key dimensions of data quality. Wang and Strong (1996) proposed a conceptual framework of four categories, which subsume 15 different data quality dimensions. These 15 dimensions of data quality are summarized in Table 4.1 together with some discussion about how data quality distortions can undermine the development and application of small area classifications.

Table 4.1 pulls together an array of challenges that combine to weaken the quality of data in developing nations. It should be stated, however, that developing countries alone do not have a monopoly of data quality problems. The data quality challenge is a universal crisis and there are examples of data quality problems reported in developed regions of the world (Schraepler and Wagner, 2005). Nevertheless, most countries in the northern hemisphere are rapidly transitioning to data-driven economies; therefore, quality data are increasingly becoming an important weapon for development. While most countries in the developed regions of the world are learning how to govern and maintain trust in their data stocks, many developing countries are not well positioned to govern data in a way that promotes development.

4.10 Mitigating Barriers

As illustrated in the Sections 4.1 to 4.9, the obstacles working against the proliferation of small area classifications in developing countries are complex and multidimensional. On the one hand, there are endogenous obstacles,

TABLE 4.1

How data quality weaknesses undermine the proliferation of small area classifications

Categories	Data Quality Dimensions	Examples of Data Quality Weaknesses
Intrinsic Data Quality	Believability	Some statutory bodies have been accused of misrepresenting demographic data records (Ferrando, 2008; Adele, 2009).
	Accuracy	The integrity of population structure reported in some major international surveys has been questioned (Randall and Coast, 2016).
	Objectivity	Political meddling and interference from interviewers during data collection lead to bias results (Randall et al., 2013; Sandefur and Glassman, 2015).
	Reputation	Repeated fabrication and deliberate manipulation of data earn statutory agencies, NGOs, and independent researchers working within these spaces a notorious reputation, which can damage credibility over time (Deaton, 2001; Devarajan, 2013; Koch-Weser, 2013).
Contextual Data Quality	Value-added	Mass deployment of emerging technologies is essential for generating new forms of measures that can give users competitive advantage and add value to their operations. The deployment of technologies such as the Internet of Things, quantum technologies, robotics, genetic engineering, spatial technologies, and artificial intelligence is growing but is still at its infancy in most developing countries (Jaiyesimi and Ojo, 2018).
	Relevancy	The collection of applicable, relevant, interesting, and usable data remains logistically difficult (Global Pulse, 2012).
	Timeliness	Vast amounts of available data are out-of-date. Consequently, precise and apt information about certain aspects of people's lives remains a mystery (UN, 2018).
	Completeness	Important characteristics about many people tend to be missing from the numbers, thereby undermining the breadth, depth, and scope of information contained in the data (Cohen, 2006; Ojo and Ojewale, 2019).
	Appropriate amount of data	In an era of data revolution, the volume of data across the developing world has also increased enormously. However, significant levels of inequalities between urban and rural areas often translate into disparities in the volumes of appropriate psychometric and behavioral data sets available for researching urban and rural dwellers. Cities in developing countries have profited more from the data revolution (Ojo and Ojewale, 2019); therefore, much less demographic data are available for studying sub-urban and rural people and places (HelpAge International, 2014).

(continued)

TABLE 4.1
(Cont.)

Categories	Data Quality Dimensions	Examples of Data Quality Weaknesses
Representational Data Quality	Interpretability	Small area classifications are increasingly dependent on the ability to fuse multiple data sets together (Leventhal, 2016). However, due to systemic and regulatory weaknesses in some developing countries, some data sets are derived from sources where key conceptual definitions may vary significantly. This can prove counterproductive when trying to elucidate results of analysis based on the combination of such data sets.
	Ease of understanding	In order to allow for ease of comprehension, it is good practice for data sets to be accompanied by metadata, which are simply data about data. Metadata are useful for resource discovery, showing how information is put together, organizing electronic resources, and facilitating interoperability (Zeng and Qin, 2008). However, due to unregulated data economies in some developing countries, some data providers often fail to provide useful metadata for their data sets.
	Representational consistency	In some countries, data administrators are confronted by poor demographic record documentation, large backlogs of data waiting to be coded, and erratic coding practices, which can make current data incompatible with previous data.
	Concise representation	Poorly designed data collection forms and database designs lead to data points that are difficult to accurately translate (Nori-Sarma et al., 2017).
Accessibility Data Quality	Accessibility	The speed at which up-to-date data can be retrieved from statutory authorities in a number of developing countries can be greatly undermined by legal challenges, bureaucratic bottlenecks, cost, and lack of technical infrastructure (Mennecke and West, 2001).
	Access security	Evidence from a range of recently developed metrics indicates that vast numbers of developing and middle-income countries are ill prepared to provide an enabling environment where access to the personal data of their citizens is shielded from illegal intrusion (Aaronson, 2019).

Source: Author's elaborations. Columns 1 and 2 (Categories and Data quality dimensions) are based on Wang, R.Y. and Strong, D.M. (1996). Beyond Accuracy: What Data Quality Means to Data Consumers. *Journal of Management Information Systems*, 12(4), 5–33.

which subsume barriers like weaknesses of institutional setup, weaknesses of organizational setup, lack of resources, lack of required infrastructure, and poor performance. On the other hand, there are also exogenous obstacles like the obstruction of data dissemination by statutory agencies, lack of comprehensive and up-to-date databases, lack of autonomy, and outdated legal frameworks. This section presents a discussion of some possible solutions to these barriers.

One of the underlying factors accounting for the credibility crisis faced by statistical agencies is a lack of professional independence. In order to assure the credibility of these statutory agencies, it is imperative to ensure that members of staff are professionally independent of political sources of interference. The promotion of political independence may also lead to better coordination of the entire national statistical system. The benefit of this is that agencies will be better able to plan and implement their activities in a participatory manner while also improving the quality, comparability, and consistency.

Some developing countries lack clear legal mandates for the collection of data. The absence of such mandates constrains the cooperation of the public. Clear laws will help to provide guidance on data collection from corporations, households, and the public. Such laws will also establish rules and disclosure controls for guaranteeing anonymity of confidential information. Responsible uses of data involve balancing political considerations and privacy concerns.

In addition to establishing clear legislations, all developing countries need to develop statistical strategies that will provide guidelines to strengthen the statistical capacity, show what and how statistics will be collected, and chart dissemination and plication approaches. Statistical strategies should also be used to identify the financial, human, and technical resources availability and requirements on the medium to long term. Statistical strategies should be demand driven, modest, and realistic and should build upon existing processes. Furthermore, mechanisms should be put in place to monitor the implementation of such strategies. In order to guarantee their effectiveness, these national statistical strategies should align with other national development strategies and budgets. This will help stakeholders from various sectors of developing economies to recognize the collection and dissemination of timely and accurate data as a thread that runs through national development agenda.

Inadequate technical and operational capacity also needs to be addressed. Greater commitment toward capacity building will ultimately reduce those barriers that obstruct the effective production and dissemination of quality statistics. Capacity building in developing countries can focus on a mix of areas. First, the continuous development of capacity for reforming underlying legal frameworks should be prioritized. There is also a need to strengthen the ability of statistical providers to properly dialogue with political stakeholders, NGOs, and users of statistics outside government.

Statistical agencies also require support in terms of their ability to adapt their organizational structures to meet the growing demand of data. Finally, specialized training and education should be regularly promoted to enable staff members update their technical capabilities.

Regular training and skills development can often be hampered by financing. Most developing countries require financial assistance in order to properly invest in infrastructure, people, and equipment. Some countries also require funding to cover short-term recurrent costs. Countries in developing regions of the world have a responsibility to increase their domestic budget allocations toward building sustainable statistical systems in the future. This may serve as an impetus for challenging donors to also raise their financial support toward a similar cause.

Data collection in some developing environments is complicated by the problems of insecurity and conflict. Data collection in conflict contexts presents several logistical challenges, including the need to cope with frequently mobile populations and the hostility that may ensue toward data collectors from local populations who feel neglected by the state. Nevertheless, numerous good practice approaches for collecting data within complex fragile environments have now been cataloged in the work of Hoogeveen and Pape (2019). Some examples of such good practice approaches include:

- Data collection using mobile phone interviews
- Rapid response surveys
- Tracking displaced persons
- Rapid consumption surveys
- Reliance on locally recruited enumerators that are trusted by their community
- Satellite images combined with sophisticated machine learning algorithms
- Video testimonials.

There is room for enhancing the quality of statistics across developing countries. To achieve this, those agencies that produce statistics within the national statistical system should be encouraged to cooperate in accordance with shared rules, principles, and standards. The generation of official statistics should be based on sound instruments, procedures, and expertise. Values that discourage impartiality while encouraging objectivity need to be promoted across national statistical agencies. Stakeholders need to commit to producing and disseminating official statistics in ways that respect scientific independence. Furthermore, better cooperation between statistical agencies in different developing countries can help enhance the exchange of knowledge and experiences.

4.11 Conclusion

This chapter has cataloged some of the major constraints working against the proliferation of small area geodemographic classifications across developing countries. It has been established that little effort has been made to promote the benefits of these classifications for developmental and humanitarian use in regions of developing countries. This is partly linked to the deep commercial roots of modern-day geodemographic practice. Widespread adoption and deployment of small area geodemographic classifications within the public sector in developed economies only commenced around the middle of the first decade of the present millennium. Consequently, it has been difficult to present good practice examples of the public policy deployment of these classifications. However, now that such public sector examples are growing in the northern hemisphere, there is an opportunity for developing regions. This can be achieved through greater and inclusive promotion of the benefits of small area classifications via traditional and nontraditional communication channels.

Much of the constraints documented here are data related. Censuses and national surveys are the foundations upon which most development data are built. Governments or external agencies run censuses and national surveys. In order to ensure the representativeness of surveys, robust knowledge about the population sampling frame is required. This knowledge can normally be derived from the population census. However, as discussed in this chapter, many countries particularly in Africa have not held a census in the last decade. This implies that large groups of the population are unaccounted for particularly within those countries that are undergoing rapid change.

Where censuses and surveys have been successfully conducted, other constraints such as compatibility, accessibility, and timeliness often surface. Nevertheless, through an appropriate mainstreaming of resources and human capital, it is possible to improve data quality and access in developing countries. The deployment of emerging technologies and the strengthening of statistical capacity should also help to ameliorate data-related challenges obstructing the proliferation of small area classifications.

References

Aaronson, S.A. (2019). *Data Is a Development Issue*. Waterloo, ON: Centre for International Governance Innovation.

Abbas, J., Ojo, A. and Orange, S. (2009). Geodemographics – A Tool for Health Intelligence. *Public Health*, 123(1), 35–39.

Adele, B.J. (2009). Falsification of Population Census Data in a Heterogeneous Nigerian State: The Fourth Republic Example. *African Journal of Political Science and International Relations*, 3(8), 311–319.

Akhmouch, A. (2012). *Water Governance in Latin America and the Caribbean: A Multi-Level Approach*. Paris: Organization for Economic Co-operation and Development.

Ashby, D.I. and Longley, P.A. (2005). Geocomputation, Geodemographics and Resource Allocation for Local Policing. *Transactions in GIS*, 9(1), 53–72.

Benedict, O.H. and Ukpere, W.I. (2012). Brain Drain and African Development: Any Possible Gain from the Drain? *African Journal of Business Management*, 6(7), 2421–2428.

Brinkerhoff, D.W. and Brinkerhoff, J.M. (2015). Public Sector Management Reform in Developing Countries: Perspectives beyond NPM Orthodoxy. *Public Administration and Development*, 35(4), 222–237.

Brown, P.J.B., Hirschfield, A.F.G., Merrall, S., Bowers, K.J., and Marsden, J. (1999). *Targeting Community Safety Projects: The Use of Geodemographics and GIS in the Identification of Priority Areas for Action*. 39th Congress of the European Regional Science Association, Dublin.

Buchmann, C. and Hannum, E. (2001). Education and Stratification in Developing Countries: A Review of Theories and Research. *Annual Review of Sociology*, 27, 77–102.

Chang, H.-J. (2011). Institutions and Economic Development: Theory, Policy and History. *Journal of Institutional Economics*, 7(4), 473–498.

Chiemeke, C.C. (2010). Bridging the Digital Divide in Developing Countries: A Case Study of Bangladesh and Kuwait. In: P. Kalantzis-Cope and K. Gherab-Martín (Eds), *Emerging Digital Spaces in Contemporary Society*. London: Palgrave Macmillan.

Cohen, B. (2006). Urbanization in Developing Countries: Current Trends, Future Projections, and Key Challenges for Sustainability. *Technology in Society*, 28(1–2), 63–80.

Dackam, R. (2003). New Strategies to Improve the Cost-Effectiveness of Census in Africa. In: UNFPA, *Counting the People, Population and Development Strategies Series*. New York, NY: United Nations Population Fund.

Deaton, A. (2001). Counting the World's Poor: Problems and Possible Solutions. *The World Bank Research Observer*, 16(2), 125–147.

Devarajan, S. (2013). Africa's Statistical Tragedy. *Review of Income and Wealth*, 59(S1), S9–S15.

DFID (2005). *Reducing Poverty by Tackling Social Exclusion*. London: Department for International Development.

Du Plessis, C. (2007). A Strategic Framework for Sustainable Construction in Developing Countries. *Construction Management and Economics*, 25(1), 67–76.

Eberhard, A. and Shkaratan, M. (2012). Powering Africa: Meeting the Financing and Reform Challenges. *Energy Policy*, 42, 9–18.

Eurostat (2013). *Guide to Statistics in European Commission Development Co-operation*. Luxembourg: Publications Office of the European Union.

Fadil, P., Williams, R., Limpaphayom, W. and Smatt, C. (2005). Equity or Equality? A Conceptual Examination of the Influence of Individualism/Collectivism on the Cross-Cultural Application of Equity Theory. *Cross Cultural Management: An International Journal*, 12(4), 17–35.

Farr, M. and Evans, A. (2005). Identifying Unknown Diabetics Using Geodemographics and Social Marketing. *Interactive Marketing*, 7(1), 47–58.

Ferrando, O. (2008). Manipulating the Census: Ethnic Minorities in the Nationalizing States of Central Asia. *Nationalities Papers*, 36(3), 489–520.

Global Pulse (2012). *Big Data for Development: Challenges and Opportunities*. New York, NY: Global Pulse.

Goodwin, S. and Dykes, J. (2012). *Visualizing Variations in Household Energy Consumption*. Poster presented at the IEEE Conference on Visual Analytics Science and Technology (VAST), 14–19 Oct 2012, Seattle, Washington, US.

Haer, R. and Becher, I. (2012). A Methodological Note on Quantitative Field Research in Conflict Zones: Get Your Hands Dirty. *International Journal of Social Research Methodology*, 15(1), 1–13.

Han, S.J., Koh, S. and Scollon, C.N. (2015). Subjective Wellbeing and Culture. In: J.D. Wright (Ed), *International Encyclopedia of the Social & Behavioral Sciences*. Amsterdam: Elsevier.

Harris, R., Sleight, P. and Webber, R. (2005). *Geodemographics, GIS and Neighborhood Targeting*. London: Wiley.

HelpAge International (2014). *The Ageing of Rural Populations: Evidence on Older Farmers in Low and Middle-Income Countries*. London: HelpAge International.

Hofstede, G. and Peterson, M. (2000). Culture: National Values and Organizational Practices. In: N. Ashkanasy, C. Wilderom and M. Peterson (Eds), *Handbook of Organizational Culture and Climate*. London: Sage Publications Ltd.

Hoogeveen, J. and Pape, U. (Eds). (2019). *Data Collection in Fragile States: Innovations from Africa and Beyond*. Cham: Palgrave Macmillan.

IEP (2018). *Global Peace Index 2018: Measuring Peace in a Complex World*. Sydney: Institute for Economics and Peace.

Indrajit, A., Van Loenen, B. and Van Oosterom, P. (2019). Assessing Spatial Information Themes in the Spatial Information Infrastructure for Participatory Urban Planning Monitoring: Indonesian Cities. *International Journal of Geo-Information*, 8(7), 305.

Jaiyesimi, R. and Ojo, A. (2018). Healthcare Planning and Emerging Technologies: Improving Healthcare Delivery. In: O. Adedeji (Ed), *Improving Life Expectancy in Sub-Saharan Africa*. Ibadan: Book Builders-Editions Africa.

Jerven, M. (2013). *Poor Numbers: How We Are Misled by African Development Statistics and What to Do About It*. Ithaca, NY: Cornell University Press.

Jimenez, E. (1995). Human and Physical Infrastructure: Public Investment and Pricing Policies in Developing Countries. In: J. Behrman and T.N. Srinivasan (Eds), *Handbook of Development Economics*. Amsterdam: Elsevier.

Kieghe, D. (2016). *National Ambition: Reconstructing Nigeria*. London: New Generation Publishing.

Kimura, Y., Saito, R., Tsujimoto, Y., Ono, Y., Nakaya, T., Shobugawa, Y., Sasaki, A., Oguma, T. and Suzuki, H. (2011). Geodemographics Profiling of Influenza A and B Virus Infections in Community Neighborhoods in Japan. *BMC Infectious Diseases*, 11(36), 1–12.

Koch-Weser, I.N. (2013). *The Reliability of China's Economic Data: An Analysis of National Output*. Washington, DC: U.S.-China Economic and Security Review Commission.

Kuehn, B.M. (2007). Global Shortage of Health Workers, Brain Drain Stress Developing Countries. *JAMA*, 298(16), 1853–1855.

Leventhal, B. (2016). *Geodemographics for Marketers: Using Location Analysis for Research and Marketing*. London: Kogan Page.

Makanga, P. and Smit, J. (2010). A Review of the Status of Spatial Data Infrastructure Implementation in Africa. *The South African Computer Journal*, 54, 18–25.

Mennecke, B.E. and West, L.A. (2001). Geographic Information Systems in Developing Countries: Issues in Data Collection, Implementation and Management. *Journal of Global Information Management*, 9(4), 45–55.

Meyer, J.-B., Charum, J., Bernal, D., Gaillard, J., Granés, J., Leon, J., Montenegro, A., Morales, A., Murcia, C., Narvaez-Berthelemot, N., Parrado, L.S. and Schlemmer, B. (1997). Turning Brain Drain into Brain Gain: The Colombian Experience of the Diaspora Option. *Science, Technology and Society*, 2(2), 285–315.

Nori-Sarma, A., Gurung, A., Azhar, G.S., Rajiva, A., Mavalankar, D., Sheffield, P. and Bell, M.L. (2017). Opportunities and Challenges in Public Health Data Collection in Southern Asia: Examples from Western India and Kathmandu Valley, Nepal. *Sustainability*, 9(7), 1106.

ODI (2017). *10 International Development Priorities for the UK*. London: Overseas Development Institute.

OECD (2015). *OECD Work on Green Growth*. Paris: Organization for Economic Co-operation and Development.

OECD (2018). *States of Fragility 2018*. Paris: Organization for Economic Co-operation and Development.

Ojo, A. and Ezepue, P.O. (2011). How Developing Countries can Derive Value from the Principles and Practice of Geodemographics, and Provide Fresh Solutions to Millennium Development Challenges. *Journal of Geography and Regional Planning*, 4(9), 505–512.

Ojo, A. and Ezepue, P. O. (2012). Modeling and Visualising the Geodemography of Poverty and Wealth Across Nigerian Local Government Areas. *The Social Sciences*, 7 (1), 145–158.

Ojo, A. and Ojewale, O. (2019). *Urbanization and Crime in Nigeria*. Cham: Palgrave Macmillan.

Ojo, E. (2010). *Freedom of Information: Current Status, Challenges and Implications for News Media*. World Press Freedom Day Conference: Freedom of Information: The Right to Know, Queensland, Australia, 3 May 2010.

Pasura, D. and Christou, A. (2018). Theorizing Black (African) Transnational Masculinities. *Men and Masculinities*, 21(4), 521–546.

Prest, S., Gazo, J. and Carment, D. (2005). *Working out Strategies for Strengthening Fragile States: The British, American and German Experience*. Paper prepared for Conference on Fragile States, Dangerous States and Failed States, University of Victoria, Victoria, BC, November 25–27.

Randall, S. and Coast E. (2016). The Quality of Demographic Data on Older Africans. *Demographic Research*, 34(5), 143–174.

Randall, S., Coast, E., Compaore, N., and Antoine, P. (2013). The Power of the Interviewer: A Qualitative Perspective on African Survey Data. *Demographic Research*, 28(27), 763–792.

Roberts, B.R. (2005). Globalization and Latin American Cities. *International Journal of Urban and Regional Research*, 29(1), 110–123.

Sabater, A. (2015). Between Flows and Places: Using Geodemographics to Explore EU Migration across Neighborhoods in Britain. *European Journal of Population*, 31(2), 207–230.

Sandefur, J. and Glassman, A. (2015). The Political Economy of Bad Data: Evidence from African Survey and Administrative Statistics. *The Journal of Development Studies*, 51(2), 116–132.

Schraepler, J. and Wagner, G.G. (2005). Characteristics and Impact of Faked Interviews in Surveys – An Analysis of Genuine Fakes in the Raw Data of SOEP. *Alleghenies Statistisches Archiv*, 89(1), 7–20.

Schwarte, C. (2008). *Access to Environmental Information in Uganda Forestry and Oil Production*. London: Foundation for International Environmental Law and Development.

Shelton, N., Birkin, M.H. and Dorling, D. (2006). Where Not to Live: A Geo-Demographic Classification of Mortality for England and Wales, 1981–2000. *Health and Place*, 12(4), 557–569.

Sietchiping, R., Permezel, M.J. and Ngomsi, C. (2012). Transport and Mobility in Sub-Saharan African Cities: An Overview of Practices, Lessons and Options for Improvements. *Cities*, 29(3), 183–189. Singleton, A.D. (2010). The Geodemographics of Educational Progression and their Implications for Widening Participation in Higher Education. *Environment and Planning A*, 42(11), 2560–2580.

Singleton, A.D. and Spielman, S.E. (2014). The Past, Present and Future of Geodemographic Research in the United States and United Kingdom. *Professional Geographer*, 66(4), 558–567.

Singleton, A.D., Wilson, A.G. and O'Brien, O. (2012). Geodemographics and Spatial Interaction: An Integrated Model for Higher Education. *Journal of Geographical Systems*, 14(2), 223–241.

Sondheim, M., Gardels, K. and Buehler, K. (1999). GIS Interoperability. In: P. Longley, M. Goodchild, D. Maguire and D. Rhind (Eds), *Geographical Information Systems: Principles and Technical Issues*. New York, NY: Wiley.

Qaisar, N., Khan, H. and Ghufran, A. (2010). E-Government Challenges in Public Sector: A Case Study of Pakistan. *International Journal of Computer Science Issues*, 7(5), 310–317.

Triandis, H.C. (2015). Collectivism and Individualism: Cultural and Psychological Concerns. In: J.D. Wright (Ed), *International Encyclopedia of the Social & Behavioral Sciences*. Amsterdam: Elsevier.

Truong, V., Dang, N., Hall, C. and Dong, X. (2015). The Internationalization of Social Marketing Research. *Journal of Social Marketing*, 5(4), 357–376.

Tumba, A.G. and Ahmad, A. (2014). Geographic Information System and Spatial Data Infrastructure: A Developing Societies' Perception. *Universal Journal of Geoscience*, 2(3), 85–92.

UN (2018). *The Sustainable Development Goals Report 2018*. New York, NY: United Nations.

UNFPA (2002). *Population and Housing Censuses: Strategies for Reducing Costs*. New York, NY: United Nations Population Fund.

Vairavamoorthy, K., Gorantiwar, S.D. and Pathirana, A. (2008). Managing Urban Water Supplies in Developing Countries – Climate Change and Water Scarcity Scenarios. *Physics and Chemistry of the Earth, Parts A/B/C*, 33(5), 330–339.

Vckovaki, A. (1998). *Interoperable and Distributed Processing in GIS*. London: Taylor & Francis.

Wang, R.Y. and Strong, D.M. (1996). Beyond Accuracy: What Data Quality Means to Data Consumers. *Journal of Management Information Systems*, 12(4), 5–33.

Webber, R. (2005). *The 2005 British General Election Campaign: How Effectively Did the Parties Apply Segmentation to Their Direct Marketing Strategies?* Paper presented at the Joint Seminar in ESRC/NERC/EPSRC Transdisciplinary Series and ESRC Research Methods Seminar Series, Center for Advanced Spatial Analysis, London.

Webber, R. and Burrows, R. (2018). *The Predictive Postcode: The Geodemographic Classification of British Society.* London: Sage Publications Ltd.

Weerakkody, V., Dwivedi, Y., Williams, M., Brooks, L. and Mwange, A. (2007). *E-Government Implementation in Zambia: Contributing Factors.* Americas Conference on Information Systems 2007 Proceedings, Keystone, Colorado, 9–12 August.

WHO (2003). *Improving Data Quality: A Guide for Developing Countries.* Manila: World Health Organization.

Willis, I., Gibin, M., Barros, J. and Webber, R.J. (2010). Applying Neighborhood Classification Systems to Natural Hazards: A Case Study of Mt. Vesuvius. *Natural Hazards,* 70(1), 1–22.

Wilkinson, R.G. (1997). *Unhealthy Societies: The Affliction of Inequality.* London: Routledge.

World Bank (2016). *Global Monitoring Report 2015/2016: Development Goals in an Era of Demographic Change.* Washington, DC: World Bank.

Zeng, M.L. and Qin, J. (2008). *Metadata.* New York, NY: Neal-Schuman.

Part 2

Underlying Techniques and Deployment Approaches

5

Building Blocks: Spatial Data Preparation

5.1 Clarifying and Defining the Purpose of Classification

Small area geodemographic classifications are ways of grouping places that are alike and different from other places in terms of key population characteristics and defining attributes of the lifestyles of the people based in these geographic areas. No one embarks upon a classification exercise without a clear reason. Researchers may have one or multiple objectives when planning to develop these small area classifications. Each objective may dictate the adaptation of specific methods or a combination of different data inputs. It is common practice for vendors of commercially driven small area classifications to suggest that their classifications are well suited for almost any purpose. However, this may not always be true as Vickers (2006) suggests that the variables or indicators used to build a particular classification may be indicative of the purpose of such classification. However, prior to discussing various reasons as to why researchers embark upon the development of small area classifications, it is essential to distinguish between two concepts, namely, general-purpose and bespoke small area classifications.

5.1.1 General-Purpose versus Bespoke Small Area Classifications

At the early stage of building small area classifications, researchers and practitioners normally need to clarify the type of area classification they wish to develop. In terms of purpose, there are two broad types of area classifications. These are general-purpose and bespoke small area classifications.

General-purpose classifications are multipurpose classifications that can be deployed within almost any sector of the economy. Such classifications are built to identify places that are clearly distinguishable from each other based on a set of core geographic and demographic characteristics that typically remains fairly stable over time, and which "makes sense" (Ojo, 2009). The advantage of general-purpose classifications is that they have multiple uses and the groups identified in such classifications are generally more stable. This makes general-purpose classifications a useful basis for change analysis.

Bespoke classifications are mainly designed to be fit for a particular purpose rather than for multiple purposes. An existing general-purpose classification may be modified to create a bespoke classification (Singleton and Longley, 2015). Alternatively, bespoke classifications may be built from scratch (Hincks et al., 2018). Different factors may influence the decision to build a bespoke classification. A researcher may feel that an existing classification does not provide the geographic granularity required for further analysis. Therefore, it may be necessary to build an area classification at the appropriate geography. There may also be a concern that a general-purpose classification subsumes out-of-date data. Alternatively, some researchers may wish to focus on a specific sector like health, education, or policing. Sector-specific classifications require specialized or niche-specific data sets which may sometimes prove challenging to secure. Bespoke small area classifications may be criticized as being narrowly focused due to their tailored uses.

For developing countries where there is an extensive paucity of small area geodemographic classifications, it makes sense to start with the development of a suite of general-purpose classifications. General-purpose classifications would provide researchers and practitioners with initial robust knowledge about the general geodemographic grouping existing within countries. Furthermore, the niche-specific nature of bespoke classifications will require specialized data sets. This may make accessing niche data sets problematic, as it is often easier to secure a mix of indicators that spread across a variety of sectors or data themes. However, it is important to stress that bespoke small area geodemographic classifications have their own values. One of the disadvantages of general-purpose classifications is that they may conceal hidden diversity within and between areas (Voas and Williamson, 2001), while bespoke classifications have the potential to exhume within and between area dissimilarities.

5.1.2 Other Purposes

It is useful to have a clear purpose for developing a classification scheme before progressing with spatial data preparation. This will be particularly useful when determining the variables or indicators that should be included in the classification. For if a classification developer intends to apply weights to variables before deploying a clustering algorithm, the weighting emphasis may be placed on those variables that critically underlie the purpose of the small area classification. This is because weights have the ability to reduce or increase the influence of particular variables on the classification (Harris et al., 2005).

Apart from general-purpose or bespoke intentions, there are other factors that influence the purpose of classifications. Gordon (1999) suggests two purposes of any area classification scheme. The first is the simplification of complex data and the second is prediction. When researchers are presented with a large data set about small areas, the development of a classification scheme represents a convenient way of organizing the complex body of data in order to enhance the efficacy of information. Small area classifications help introduce some simplicity and structure to a potentially endlessly complex picture of differences within and between areas.

Another motivation for the creation of an area classification might be for that classification to function as a precursor to a more complex classification at a finer geographic detail. For instance, Vickers (2006) had the intention of classifying 223,060 UK output areas. However, he concluded that starting out with an output area classification might present challenges due to the volume of data. Using the same census data sets at his disposal, he decided to build an initial classification of 434 local authorities of the UK. This pilot area classification of local authorities enabled him to better understand his data, test various techniques, and learn lessons, which proved useful during the development of a more complex classification of output areas.

The motive behind developing an area classification might be to enable a shared understanding and language about complex phenomena by introducing simplicity and patterning (Johnston, 1976). Similarly, some researchers develop area classifications with the intention of gauging likely demands for particular policy products and services.

The classification of small areas can also be a tool for data exploration, particularly if such a classification is based on a set of descriptive statistics. For instance, if the intention is to explore the geographic distribution of a single indicator, then that indicator may be mapped into classes above or beneath the national average. For example, in the Philippines, *barangay* is the native Filipino term for a village, district, or ward. It represents the smallest administrative division in the country. To generate Figure 5.1, the mean population density of all *barangays* was calculated and those areas that are above and beneath the average are differentiated.

In order to further isolate and explore those *barangays* that are very high above or beneath the national average, a standard deviation technique could be used to categorize population density as shown in Figure 5.2. It should be noted however that the data exploration examples shown in Figures 5.1 and 5.2 are based on single variables. Small area geodemographic classifications are particularly useful for exploring multivariate data.

5.2 Principles for Selecting Initial Input Variables

The development of small area geodemographic classifications focuses on grouping objects (i.e., small geographic areas) into similar categories based on multidimensional attributes of people and the characteristics of their environment. Once a researcher or a practitioner has clarity about why he or

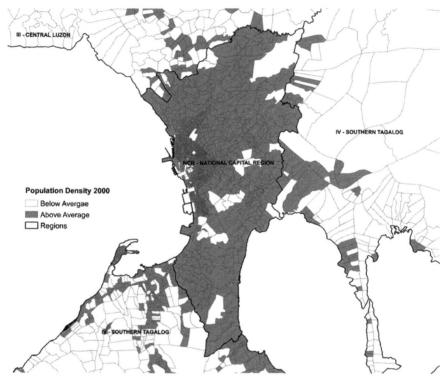

FIGURE 5.1
A simple manually adjusted classification scheme for univariate data.

she wishes to develop a small area classification, the next task is to assemble attributes that correspond to the administrative geography (also called geographic scale), for which the classification scheme will be developed. These attributes are derived from variables.

A variable is a characteristic of an object being observed that may assume any value among a set of values to which a numerical measure or a category can be assigned (Dodge, 2003). Essentially, for each small area that will be classified, measurements will be taken, for example, on variables such as age, income, and occupation. It is good practice to select variables for inclusion if there is a cogent reason for their inclusion. There are different viewpoints on what constitutes a good practice when selecting input variables for the development of a small area classification. Figure 5.3 illustrates some of the key principles that may be borne in mind during the initial selection of input variables.

5.2.1 Theoretical Relevance

There is usually a temptation to dive straight into selecting input variables without considering their theoretical relevance for the classification exercise.

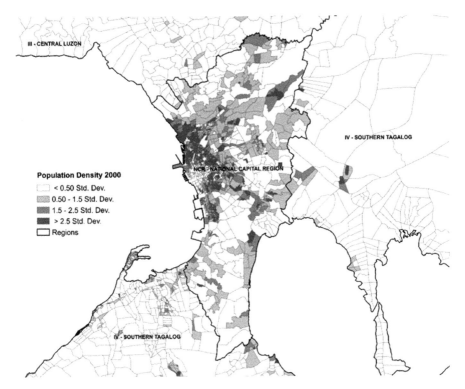

FIGURE 5.2
A simple standard deviation classification scheme for univariate data.

This happens when there is a lack of theoretical framework underpinning the entire classificatory process. There are good reasons as to why theory should serve as the point of departure when selecting initial input variables. Variables that are not theoretically informed are unlikely to be grounded in the existing body of knowledge. The inclusion of such variables may also be to generate results of a narrow and ungeneralizable value (Yiannakis, 1992). Furthermore, in order to operationalize variables, they must first be defined before they can be constructed. This can be achieved by linking a relevant theory with concepts and related observations.

5.2.2 Objectivity

To understand the principle of objectivity, it is necessary to distinguish between objective and subjective measures. Objective measures are those that involve an impartial measurement, while subjective measures are influenced by the observer's personal judgment and skill (Schachter, 2010). It is desirable to ensure that small area classifications subsume variables that are well defined and have a clear meaning. Objective variables will be precise and

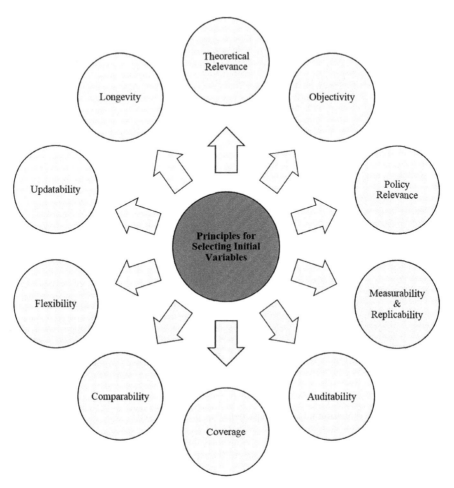

FIGURE 5.3
Variable selection principles.

they are necessary to avoid the trap of ambiguity when describing groups or categories that emerge after developing a small area classification. These variables are also easy to understand and they help classification developers to circumvent the traps of bias and prejudice.

5.2.3 Policy Relevance

The development of small area classifications in developing countries needs to be informed by policy relevance in order to attract the interest of policymakers. Consequently, variables should be selected within the context of local, regional, and possibly global policy agendas. For instance, variables that measure issues connected to poverty directly or indirectly often motivate action from the local, state, and national governments. Such issues could be borne in mind when making a choice of variables for the development of small area classifications.

5.2.4 Measurability and Replicability

The ability to quantify and reproduce variables should also inform the selection procedure. Measurable replicable variables are systematically observable, and they have the capacity to be analyzed and tested. Measurability and replicability are principles that help embellish the ethos of research transparency in the entire development process of small area classifications.

5.2.5 Auditability

Auditability is a principle that also seeks to promote the ethos of research transparency. Variables included in classifications should be generated in ways that allow for scrutiny and third-party validation. Auditability is also linked to the principles of measurability and replicability. If variables cannot be measured and reproduced by other researchers, how will the results emanating from that research study be validated or challenged if necessary? The more auditable variables are, the more credible and legitimate an area classification is.

5.2.6 Coverage

Variables with wide geographic coverage should be prioritized and they have the tendency to improve statistical significance. Census variables have a benefit of wide geographic coverage but face the problem of updatability. In some developing countries, censuses are not conducted in regular intervals. This raises the case for considering variables derived from national surveys. Many of the current widely used classifications in the developed world supplement census statistics with data derived from lifestyle surveys (Harris et al., 2005). The argument is that such surveys are conducted much more frequently; they utilize the same geographies as the census and ask more direct questions beyond population demographics and housing.

5.2.7 Comparability

Another useful selection principle is the comparability of variables. Comparability is about ensuring coherence and consistency. Comparable variables are those variables that have been standardized over time (longitudinally) and across geographies (transversally). Those variables that have been generated and compiled based on common standards (scope, definitions, and units) promote the principle of comparability. It is recognized that this principle can be difficult to imbibe especially where data used to produce small area classifications are derived from multiple sources. Nevertheless, it is recommended that variables from different data sources and of different periodicity should be compared and reconciled.

5.2.8 Flexibility

As new evidence emerges, definitions and concepts underlying social research practices may be subjected to modifications over time. Similarly, the emergence of new evidence about the effectiveness of policy interventions may necessitate the need to adjust how things are defined or measured. Modifications of conceptual interpretations often trigger a corresponding requirement to adjust existing variables. Useful variables should therefore be flexible and capable of accommodating continuous improvements to what is measured and how it is measured.

5.2.9 Updatability

When selecting input variables, researchers should also prioritize those variables that have the potential to be updated continuously in the future. The use of updatable variables will give room for the development of small area classifications that can be used to analyze social change, for instance (Vickers and Rees, 2007; Gale et al., 2016). The incorporation of updatable indicators can also help ensure that small area classifications are deployed as long-term policy evaluation frameworks.

5.2.10 Longevity

Classification developers will not want to develop area classifications that are out of date as soon as they have been made. Therefore, variables that can sustain the life span of the small area classification system are highly desirable. One key theoretical principle that should be considered during the choice of variables is the sensitivity of each variable to change. The rate at which the values of variables change over time can have implications on the durability and reliability of the area classification (Vickers, 2006). Variables that will sustain the classification over its life course are those that do not have the potential of yielding to large changes across areas.

5.3 Quality Control and Reduction of Initial Variables

The assembling of an initial pool of variables based on the principles discussed in Section 5.2 may normally generate a large pool of arbitrary variables. At this initial stage, little attention is given to the interrelationships between variables. It is important to ensure that the most relevant variables are included in the development process of building small area classifications (Harris et al., 2005). This implies that not all the variables available at the initial stage would be appropriate inputs.

Another important reason as to why it is essential to ensure that variables included in an area classification are carefully selected is that the proposed variable needs to reflect real-life social patterns. Implementing a critical and statistically rigorous variable selection approach can help ensure that the area classification is robust.

Different opinions have been expressed about the size of variables that should be used to create small area classifications. Some researchers have argued for fewer variables (Voas and Williamson, 2001), while others, particularly those who seem to favor commercially driven small area geodemographic classifications, support the view that the more the variables, the more robust the system is likely to be (Harris et al., 2005; Leventhal, 2016). For developing countries where small area classifications are rare, these debates generate more questions than answers. For instance, what is few and what is more? There is no reference material that states the standard or specific number of variables that should be used to develop an area classification. Therefore, the focus should be to ensure that the final variables chosen are conceptually appropriate and analytically relevant. In the remainder of this section, a number of multivariate statistical techniques for judging the appropriateness of input variables are discussed. These techniques can be used to evaluate the underlying structure of data, identify groups of variables that are statistically similar, and match the statistically determined structure of the data to the theoretical framework that underlies the intended small area classification.

5.3.1 Principal Components Analysis

One of the key reasons as to why it is important to carefully examine an initial pool of variables and reduce them before further analysis is because of the possible presence of multicollinearity within the data set. Multicollinearity is a situation where two or more variables are highly correlated (Walker and Maddan, 2008). It is a feature that is quite common within sociodemographic variables. Multicollinearity is not desirable when clustering variables together because it duplicates information. A preliminary data preparatory analysis is needed to determine which variables represent the most the principal dimensions of an original data set.

Principal components analysis (PCA) is a common technique that is used to achieve this task. It is an analytical procedure used to alter the relationship of correlations existing within a data set by transforming a set of correlated variables into a smaller set of uncorrelated variables (Jolliffe, 2002). The PCA technique works by analyzing and explaining the variance within an observed data set by linearly combining the original data. These linear relations are termed principal components of the data as illustrated in the notations in Equation 5.1.

$$Z_1 = a_{11}x_1 + a_{12}x_2 + \ldots a_{1Q}x_Q$$

$$Z_2 = a_{21}x_1 + a_{22}x_2 + \ldots a_{2Q}x_Q$$

$$Z_Q = a_{Q1}x_1 + a_{Q2}x_2 + \ldots a_{QQ}x_Q \qquad (5.1)$$

where :

x_1, x_2 and x_Q represent the variables in the data set

Z_1, Z_2 and Z_Q represent the uncorrelated principal components, or linear relations of the original data.

What this means is that while PCA in cluster analysis can help mitigate the problem of data redundancy (Everitt et al., 2001; Harris et al., 2005), the new data values which it creates (called composites) may not be easy to explain or interpret if included in the cluster analysis.

Another usefulness of PCA as a data reduction technique is its ability to identify those variables that are likely to have significant influence on the clustering process (Jolliffe, 2002). One of the outputs of the analysis is a weights table, also called a components loading matrix. This matrix is indicative of the power that each variable exerts. Typically, for a variable, the higher the value of this weight, the better its variance can be explained by the corresponding principal component, and consequently the greater the power it is likely to exert on the cluster analysis (Jolliffe, 2002).

While using PCA to determine those variables that are likely to power a data set, it is advisable to examine different components in order to see how variables behave in terms of the magnitude of their loading. Usually, variables with higher loading values are more likely to have greater influence on the data set (Vickers, 2006). A trade-off however needs to be made between the different components to decide which one will inform the choice of variables that will most likely power the classification. Since the first principal component usually accounts for the greatest level of variation within the data (Dunteman, 1989), it is safe to use this component to determine those variables that are likely to have the greatest influence on the classification.

5.3.2 Missing Values, Small Sample Sizes, and Creating Composite Variables

Apart from using census data sets, which have nationwide coverage, small area geodemographic classifications are often supplemented with survey data sets. Although data from surveys have the potential of supplying information not contained in national censuses (Harris et al., 2005), one of the shortcomings suffered by such data sets is that they often contain missing values for certain geographic areas (Ojo and Ezepue, 2011). Where a variable contains too many missing values, it provides incomplete information, which is not too helpful for further analysis.

Another important issue to consider when choosing the final list of input variables is the proportion of the population represented by a variable (Vickers, 2006). The fact that a variable is derived from the national census does not necessarily mean that it will represent a large percentage of the population. When a variable represents only a small proportion of the population, there is the tendency for that variable to be volatile and change rapidly over time. Such a variable may not sustain the longevity of the classification (Vickers and Rees, 2007).

Similarly, those variables that have small sample proportions provide little distinctive information for naming and profiling the groups created during the cluster analysis (Harris et al., 2005). A solution proposed for solving this problem is to merge the variables if they fall under the same theme and share a similar base. For instance, rather than using single year age variables, these can be merged into composite variables of five-year age groups (0 to 4 years, 5 to 9 years, etc.) in order to make them more stable and robust. Apart from helping to deal with the problem of small sample sizes, the creation of composite variables also helps reduce the large number of initial variables considered for inclusion in a cluster analysis.

When attempting to merge two or more variables, it is essential to ensure that the variables share the same denominator. However, it is also imperative to ensure that the merger of variables makes sense. For instance, women in developing countries receive assistance from different people during the delivery of their children. These include doctors, nurses, midwives, auxiliary midwives, traditional birth attendants (TBAs), relatives, and friends. Some women do not receive support from anyone during child delivery. It makes sense to combine the different types of assistance received during childbirth into two composite variables: skilled and unskilled assistance. The composite variable "skilled assistance" would subsume doctors, nurses, midwives, and auxiliary midwives, while the composite variable "unskilled assistance" would subsume TBAs, relatives, friends, and no assistance.

5.3.3 Internal Consistency and Reliability of Variables

Generally, many concepts of interest in the social sciences are difficult to measure explicitly. In such cases, a series of questions are typically asked through census and survey data and answers are combined into a single numerical value. These procedures present internal consistency and reliability challenges – that is, how well a set of variables or items measures a single, one-dimensional concealed aspect of individuals in a population. Internal consistency and reliability are important for this analysis because in their absence, it is impossible to have any validity associated with the outputs generated from the analysis. It is therefore useful to assess the internal consistency of variables prior to selecting a final list for the clustering process.

The Cronbach's coefficient alpha (C-alpha) (Cronbach, 1951) given by the notation in Equation 5.2 is the most common estimate of internal consistency of data (Raykov, 1998; Boscarino et al., 2004). It assesses how well a set of indicators measures a single one-dimensional object.

$$\alpha_c = \left(\frac{Q}{Q-1}\right)\frac{\sum_{i \neq j} cov\left(x_i, x_j\right)}{var\left(x_0\right)}$$

$$= \left(\frac{Q}{Q-1}\right)\left[1 - \frac{\sum_j var\left(x_j\right)}{var\left(x_o\right)}\right]$$

$$c = 1,\ldots M; j = 1,\ldots Q \tag{5.2}$$

where:

M represents the number of geographic areas considered

Q represents the number of variables available

$x_0 = \sum_{Q=1}^{Q} x_j$ represents the sum of all individual variables.

The C-alpha measures the portion of total variability of the sample of individual variables due to the correlation of indicators. To interpret the C-alpha statistic, a commonly accepted rule of thumb proposed by George and Mallery (2003) could be adopted for describing internal consistency. This rule is illustrated in Table 5.1. The C-alpha increases with the number of individual variables and with the covariance of each pair. If no correlation exists and individual variables are independent, then C-alpha is equal to zero; if individual variables are perfectly correlated, then C-alpha is equal to one. C-alpha should not be interpreted as a statistical test, but a coefficient of

TABLE 5.1

Rule for interpreting Cronbach's Alpha

C-Alpha Value	Internal Consistency
$\alpha_c \geq 0.9$	Excellent
$0.9 > \alpha_c \geq 0.8$	Good
$0.8 > \alpha_c \geq 0.7$	Acceptable
$0.7 > \alpha_c \geq 0.6$	Questionable
$0.6 > \alpha_c \geq 0.5$	Poor
$0.5 > \alpha_c$	Unacceptable

reliability based on the correlation between individual indicators. That is, if the correlation is high, then there is evidence that the individual indicators measure the same underlying construct. Therefore, a high C-alpha, or equivalently a high "reliability", indicates that the individual indicators measure the latent phenomenon well.

5.3.4 Issues Relating to Skewness

Prior to choosing a variable for inclusion in a clustering algorithm, it is useful to consider the extent to which it is skewed. Skewness is the extent to which a variable is symmetrically distributed about its mean (Crawshaw and Chambers, 2001). Three forms of skew are illustrated in Figure 5.4.

A variable exhibits positive skew if its asymmetric tail (when charted) extends toward the positive values of the distribution. For negative skew, the tail extends toward the negative values. A normally distributed variable would produce a skewness value of zero, which would ideally be desirable for variables to be included in the cluster analysis (Harris et al., 2005). However, in practice, this is rarely the case with spatially referenced sociodemographic data sets.

$$G_1 = \frac{\sum_{i=1}^{N} \left(Y_i - \bar{Y}\right)^3 / N}{S^3} \qquad (5.3)$$

where:

G_1 represents the value of skew
$\bar{Y}$ represents the mean of the distribution
S represents the standard deviation of the distribution
N represents the number of data points.

The Fisher–Pearson coefficient of skewness given in Equation 5.3 can be used to evaluate the degree of skewness of each variable within a data set (George and Mallery, 2003). Others have also found Shapiro–Wilk and

FIGURE 5.4
Illustration of negative skew, normal distribution and positive skew.

Kolmogorov–Smirnov tests as useful for evaluating nonnormal distributions (Field, 2009).

Skew can result from a number of issues. If a variable represents a small proportion of the population, most values will be concentrated around the lower end of a 0 to 100% scale. Skew can also be caused by the presence of extreme values or outliers in a data set. An example of a variable that may demonstrate this feature in most developing countries is population density. In most developing countries where urbanization is rapid, cities tend to have disproportionately higher population concentrations, while rural areas generally exhibit lower densities of people (Ojo and Ojewale, 2019). The problem with introducing highly skewed variables into the analysis is that such variables may obscure the rest of the data set and create artificial groups after the classification exercise. It is therefore important to test variables for their skewness and avoid where possible the inclusion of highly skewed variables. In practice, however, it is practically impossible to eliminate the presence of some skewed variables when creating small area classifications. Their effects can be minimized by transforming the data (Walker and Maddan, 2008). Methods of transformation are discussed in Section 5.4 of this chapter.

5.3.5 Cross-Correlation and Geographic Dispersion

Usually, a pool of independent variables is considered for inclusion in a cluster analysis. The fact that variables are independent of each other does not necessarily mean that certain types of association do not exist among them. It is important to determine the types of relationship that exist between variables before including them in cluster analysis.

The knowledge of the relationship existing between two or more variables can be indicative of the level of redundancy that may be introduced into a small area classification (Walker and Maddan, 2008). Furthermore, the inclusion of two highly related variables in a clustering algorithm will often result in the repetition of the same population characteristic (Milligan and Stephen, 2003). This can give undue advantage to certain variables and also mask other important underlying characteristics existing within the population. In mathematical terms, the level of association between two or more variables can be determined using the coefficient of correlation. Although there are various measures of correlation, the most common metric is the Pearson product moment correlation (PPMC) (Crawshaw and Chambers, 2001).

Three types of correlations tend to be visible in sociodemographic data assembled for clustering. The first type of correlation will be present within those pairs of variables that share the same denominator. Naturally, these variables tend to exhibit negative correlations. The second type of correlation emerges because of causality. This suggests that one or more variables may be fundamental properties of others even though they do not share the same denominator. The third type of correlation may be detected where the presence of one variable implies the presence or absence of another variable.

TABLE 5.2
Effects of composite variables on the standard deviation

Geographic Area	Var 1 [Sometimes Has Food Needs (%)]	Var 2 [Often Has Food Needs (%)]	Var 3 [Always Has Food Needs (%)]	Composite Var 1, 2 and 3 [Food Needs (%)]	Composite Var 2 and 3 [Food Needs (%)]
A	59.35	2.11	0.66	62.12	2.77
B	56.99	16.50	0.00	73.48	16.50
C	50.25	7.72	5.28	63.25	13.00
D	48.74	8.03	0.93	57.71	8.97
E	47.68	6.50	15.60	69.77	22.10
F	46.96	9.30	14.55	70.82	23.86
G	44.91	2.43	20.15	67.49	22.58
H	31.20	6.37	23.23	60.80	29.60
I	24.33	9.33	25.57	59.23	34.90
J	38.53	4.16	25.29	67.98	29.45
Standard Deviation	**10.86**	**4.16**	**10.54**	**5.36**	**10.04**

For variables to work in the classification they need to show variation over geographic space. A useful statistic for measuring the geographic variation of variables is the standard deviation. The standard deviation of a data set is a measure of how spread out the data are. Essentially, it is calculated by measuring the average distance from the mean of the data set. Variables with larger standard deviations will prove more useful than those with lower values (Harris et al., 2005). This is because they present better distinctions between areas. However, it is important to consider the sample sizes of variables across areas especially when dealing with variables with low standard deviation values. Some of these variables can be merged into composite variables in order to increase their standard deviations. However, care must be taken when merging variables as it could also prove counterproductive as shown in Table 5.2. In the example presented, variables 1, 2, and 3 show the percentage of the population with different levels of food needs. The merger of all the three variables reduces the geographic variation of the composite variable (column 5) unlike the merger of only variables 2 and 3 (column 6).

5.4 Dealing with Outliers and the Problem of Different Measurement Units

The next two stages in the preprocessing of spatial data involve resolving the problems of skew and the challenge posed by different measurement units. Highly skewed variables can be adjusted through the process of transformation (Harris et al., 2005). Some key transformation methods are discussed in this section. Similarly, the assembled variables may be available at different measurement scales and units, such as percentages, rates, ratios, and kilograms, etc. It is inappropriate to run a machine learning clustering algorithm on a data set comprising variables that have been measured using different units. Doing this would give undue advantage to certain variables at the cost of others. Normalization and standardization techniques help in rescaling variables (Berrar et al., 2003).

5.4.1 Data Transformation

One of the key contributors to highly skewed data distributions can be the presence of numerous outliers. Outliers are simply extreme values or data point that behave very differently from the rest of the data distribution. Transformation is a procedure for altering data distributions in order to reduce the effect of outliers. It can be useful to transform data prior to normalization or standardization although transformation is not a compulsory data preprocessing task. Vickers (2006) suggests that the effectiveness of transformation techniques needs to be evaluated in two ways:

- The extent to which the transformation improves the overall distribution of the data set
- The extent to which the transformation affects or distorts the integrity of the data set.

Positive skew is the most common type of skew found in sociodemographic data sets. Some have found the log transformation to be particularly helpful when trying to resolve this problem (Vickers, 2006).

$$b_{ij} = log\left(x_{ij}\right) \tag{5.4}$$

where:

x_{ij} represents the original value of the variable in row i and column j

b_{ij} represents the adjusted value of the variable.

As shown in Equation 5.4, the logarithm transformation works by relating a base number to an exponent. The exponent to which a chosen base value must be raised in order to derive the original value of the variable is the logarithm. When this method is used on data, natural logarithms of the variable are used to run the clustering algorithm rather than the original values of the variable.

$$b_{ij} = log\left(x_{ij} + 1\right) \tag{5.5}$$

where:

x_{ij} represents the original value of the variable in row i and column j

b_{ij} represents the adjusted value of the variable.

Some variables may contain zeros. To log-transform variables containing zeros, it is common practice to add a number to all the data points as shown in the notation given in Equation 5.5.

Other forms of transformation, which are not as popular as the log transformation may also be considered. The square root transformation, which involves taking the square root of all values in a distribution, can also be considered (Crawshaw and Chambers, 2001). One thing to bear in mind with this method is that it is impossible to calculate the square root of a negative value. If a distribution has negative values, then a constant will need to be added to all values to move them above the zero mark.

Reciprocal transformation involves dividing 1 by the original value of a variable. The resultant value is an inverse of the original value of the variable. Reciprocal transformations have a characteristic of making very small numbers large and vice versa, thereby reversing the order of the original data. This attribute may not be too helpful for a cluster analysis procedure.

Another way to handle outliers is to cap them. For instance, if one finds that people above a certain income level behave in the same way as those with a lower income, then in such scenarios the income variable may be capped at a level that keeps the distribution intact. Vickers (2006) has also suggested capping extreme values data at various levels, for instance, at the top and bottom 1% in order to reduce the effect of outliers.

5.4.2 Normalization and Standardization

It is inappropriate to combine variables that are measured in different units because the resultant output will not be a true reflection of reality. Variables measured on a unit with a large range may be given undue advantage when compared to those measured on a smaller range (Harris et al., 2005). Normalization and standardization techniques can be deployed on data to ensure that they are in comparable units.

Normalization procedures are used to rescale variable values without necessarily ensuring that the mean is equal to zero or that the standard deviation is equal to one (Angelov and Gu, 2019). Some normalization methods are used to cap the normalized variable within a specific range, for instance, between zero and one. In addition, if the intention is to ensure that all values are positive, then a normalization method such as the range normalization may be the preferred choice for rescaling variables. However, some other normalization methods, such as inter-decile and interquartile range normalization, will generate a mix of negative and positive values.

One of the popular normalization approaches utilizes the minimum and maximum values of a variable distribution to rescale data between a minimum value of zero and a maximum value of one. An example is the range normalization shown in Equation 5.6, which works by subtracting the minimum value and dividing by the range of the variable distribution (Wallace and Denham, 1996; Vickers, 2006). This method has the advantage of working relatively well with extreme values by shrinking all values within limited range. However, the shrinking of extreme values could also distort the distribution of the transformed variable.

$$S_i = \frac{x_i - x_{min}}{x_{max} - x_{min}} \qquad\qquad (5.6)$$

where:

S_i represents the normalized value of the variable for area i

x_i represents the original value of the variable for area i

x_{min} represents the minimum value of the variable across all areas

x_{max} represents the maximum value of the variable across all areas.

Another normalization method for rescaling data distributions works with the inter-decile range. The difference between the first and ninth deciles is known as the inter-decile range (Kirch, 2008). This method is a slight variation of Wallace and Denham's (1996) range normalization method. While the range normalization is calculated by comparing the minimum and maximum values of a distribution, the inter-decile range approach is calculated by relating the median and 10th and 90th percentiles of a distribution as shown in the notation given in Equation 5.7.

$$S_i = \frac{x_i - x_{med}}{x_{90^{th}} - x_{10^{th}}} \qquad\qquad (5.7)$$

where:

S_i represents the normalized value of the variable for area i

x_i represents the original value of the variable for area i

x_{med} represents the median of the variable across all areas

$x_{90^{th}}$ represents the 90th percentile

$x_{10^{th}}$ represents the 10th percentile.

A third normalization technique is the interquartile range normalization shown in Equation 5.8. The interquartile range is the difference between the first and third quartiles (Qin et al., 2012). The first quartile is the value in the data set that holds 25% of the values below it. The third quartile is the value in the data set that holds 25% of the values above it.

$$S_i = \frac{x_i - x_{med}}{Q_3 - Q_1} \tag{5.8}$$

where:

S_i represents the normalized value of the variable for area i

x_i represents the original value of the variable for area i

x_{med} represents the median of the variable across all areas

Q_3 represents the third quartile

Q_1 represents the first quartile.

Interquartile range normalization is computed by dividing the difference between the original value of the variable and the median value of the distribution by the difference between the upper and lower quartiles.

In multivariate analysis, standardization or z-scores work slightly differently. This method is used to rescale variables to a new distribution that has a mean of zero) and a standard deviation of one (Crawshaw and Chambers, 2001). Standardization is one of the most common methods used to convert variables into a common scale. If this method is used, variables with extreme values could have a greater effect on the eventual small area classification typology.

$$Z_i = \frac{x_i - \mu}{\sigma} \tag{5.9}$$

where:

Z_i represents the standardized value of the variable for area i

x_i represents the original value of the variable for area i

μ represents the mean value of the variable across all areas

σ represents the standard deviation of the variable across all areas.

Equation 5.9 shows how z-scores are computed by dividing the difference between the original value of the variable and the mean across all geographic areas by the standard deviation of the variable across all areas.

5.5 Weighting Techniques

Adding a weight to a variable signifies the level of importance of such a variable (Gnanadesikan et al., 2005) to an area classification system. However, a variable that is considered important for one purpose may not be important for another purpose. Furthermore, if certain variables are deemed more important within a data set, then there may be differences in the levels of importance associated with such variables.

Some of the factors that influence the decision to weight variables include the knowledge of the context within which the small area classification is being built. Judgments about weighting variables can also be influenced by local policies. For instance, national development plans often contain short- to long-term policy priorities that may give an indication of those issues that are more or less important to policy stakeholders (Ojo et al., 2018). Public sentiment could also have an influence on the decision to assign weights to variables.

Another reason as to why weights are used in the preprocessing of spatial data prior to cluster analysis is to compensate for the quality of the statistical data set available for analysis (Gnanadesikan, 2007). For instance, variables may be compensated due to their geographic coverage. Care must be taken to ensure that some variables are not penalized for others.

Since the importance associated with variables can be a source of contention, several scholars agree that it is not compulsory to weight input variables before deploying a machine-learning clustering algorithm (Gordon, 1999; Everitt et al., 2001; Harris et al., 2005). Indeed, during the entire variable selection process, some form of unintended or subjective weighting occurs. Choosing one variable over another means that in principle the chosen variable is considered more important than the excluded variable.

If a researcher is not completely convinced about the theoretical and statistical importance of variables within a data set, then it is safer to apply equal weights (or no weighting) to the variables. However, when working with equal weights, it is important to ensure that multicollinearity is reduced to the barest minimum. Multicollinearity is the occurrence of high intercorrelations among independent variables (Alin, 2010). If two or more collinear variables are included in the analysis, then the spatial dimension that they represent may be doubled or tripled.

Variable weighting decisions can also be made based on the valuations of experts in specific fields. For instance, health experts may be approached to give their opinions on specific health variables that may be incorporated into a small area classification system. Similarly, education, gender, and public policy professionals can be consulted to give advice about the importance of specific variables based on their experience. Weighting data based on the expert advice of professionals is a subjective process, which may affect the degree of acceptance of the baseline weighting scheme.

Factor analysis can also be used as a method for obtaining weights for variables. This method was used to weigh variables during the development of a multivariate City Infrastructure Quality Index for Nigeria (Ojo et al., 2018). The underlying conjecture for the generation of weights with factor analysis is the assumption of the existence of latent constructs of the policy areas.

Factor analysis is used to describe a data set with a set of orthogonal factors, which are considerably large. This method assumes that variables measure the underlying construct with varying degrees of accuracy. Therefore, variables that do not correlate highly with the common factor (i.e. those with a low common factor loading) are given a lower weight in the construction of the factor score (Rummel, 1967).

$$X_k - \mu_k = l_{k1}F_1 + l_{k2}F_2 + ... l_{km}F_m + \varepsilon_k, k = 1,2,...,p \tag{5.10}$$

where:

p	represents the observable variables in the data set
m	represents unobservable random components
$F_1, F_2, ..., F_m$	represent the common factors
$\varepsilon_1, \varepsilon_2, ... \varepsilon_p$	represent additional sources of variation specific to a variable.

The weight l_{k1} is known as the loading of the kth variable on the jth factor. The portion of the variance of the kth variable contributed by the m common factors is known as the kth communality. It is the sum of squared factor loadings (Fabrigar et al., 1999).

Two factor extraction procedures are associated with factor analysis: principal axis factoring (PAF) and maximum likelihood estimation (MLE). When deciding on the extraction procedure to adopt, it is important to bear in mind that the PAF is better able to recover weak factors while the MLE estimator is asymptotically efficient (i.e. it may take fewer computational steps). Furthermore, it has been argued that if data are relatively normally distributed, the MLE should be the preferred choice because "it allows for the computation of a wide range of indexes of the goodness of fit of the model and permits statistical significance testing of factor loadings and correlations among factors and the computation of confidence intervals" (Fabrigar et al., 1999, p. 277). However, if the assumption of multivariate normality is severely violated, the PAF is recommended (Fabrigar et al., 1999).

5.6 Conclusion

The process of organizing observable data into groups takes place by clustering the data into meaningful structures. Some researchers embark upon the clustering of geographic data using machine-learning algorithms without carefully preprocessing the data set. This is dangerous because passing any data set through a clustering algorithm will generate groups. However, the real question is how representative of reality will be the emerging groups. This is why it is essential to preprocess the data prior to clustering them. Kaufman and Rousseeuw (2009) have suggested that the clustering elements (which are the geographic areas to be clustered) should, where possible, be defined to give a 100% geographic coverage. It is also suggested that these objects should be representative of the cluster structure believed to be present if this is known. The attributes of the clustering objects will be defined by variables, which represent measurements taken on each geographic area. Most commercial providers of small area classifications often suggest that the greater the number of variables included in an area classification, the better it is. This view contrasts sharply with evidence-based positions expressed in the academic literature, which suggest otherwise. Variables should be selected based on their analytical soundness, measurability, coverage, relevance to the phenomenon being measured, and relationship with each other. Where data are scarce, proxy variables can be considered. There is no specific requirement that transformation must be performed on any set of data. However, it can be used to reduce the effect of skew. Normalization and standardization are more likely to be adapted, as most variables are likely to be a mix of measurements computed using different units. In the unlikely event that the same units are used for measuring all variables, normalization and standardization may be ignored. It is up to the researcher to decide if weighting is necessary and if so which method should be used.

References

Alin, A. (2010). Multicollinearity. *WIREs Computational Statistics*, 2(3), 370–374.

Angelov, P.P. and Gu, X. (2019). *Empirical Approaches to Machine Learning*. Switzerland, Cham: Springer Nature.

Berrar, D.P., Dubitzky, W. and Granzow, M. (Eds). (2003). *A Practical Approach to Microarray Data Analysis*. New York, NY: Kluwer Academic Publishers.

Boscarino, J.A., Figley, C.R. and Adams, R.E. (2004). Compassion Fatigue Following the September 11 Terrorist Attacks: A Study of Secondary Trauma among New York City Social Workers. *International Journal of Emergency Mental Health*, 6(2), 1–10.

Crawshaw, J. and Chambers, J. (2001). *A Concise Course in Advanced Level Statistics with Worked Examples*. Cheltenham: Nelson Thornes.

Cronbach, L.J. (1951). Coefficient Alpha and the Internal Structure of Tests. *Psychometrika*, 16, 297–334.

Dodge, Y. (Ed). (2003). *The Oxford Dictionary of Statistical Terms*. Oxford: Oxford University Press.

Dunteman, G.H. (1989). *Principal Components Analysis*. Newbury Park, CA: Sage Publications.

Everitt, B.S., Landau, S. and Leese M. (2001). *Cluster Analysis*. London: Arnold.

Fabrigar, L.R., Wegener, T., MacCallum, R.C. and Strahan, E.J. (1999). Evaluating the Use of Exploratory Factor Analysis in Psychological Research. *Psychological Methods*, 4(3), 272–299.

Field, A. (2009). *Discovering Statistics Using SPSS*. London: Sage Publications.

Gale, C.G., Singleton, A.D., Bates, A.G. and Longley, P.A. (2016). Creating the 2011 Area Classification for Output Areas (2011 OAC). *Journal of Spatial Information Science*, 12, 1–27.

Gnanadesikan, R., Kettenring, J.R., and Srinivas, M. (2007). Better Alternatives to Current Methods of Scaling and Weighting Data for Cluster Analysis. *Journal of Statistical Planning and Inference*, 137(11), 3483–3496.

George, D. and Mallery, P. (2003). *SPSS for Windows Step by Step: A Simple Guide and Reference*. Boston, MA: Allyn and Bacon.

Gnanadesikan, R., Kettenring, J.R and Tsao, S.L. (2005). Weighting and Selection of Variables for Cluster Analysis. *Journal of Classification*, 12(1), 113–136.

Gordon, A.D. (1999). *Classification*. London: Chapman and Hall.

Harris, R., Sleight, P. and Webber, R. (2005). *Geodemographics, GIS and Neighborhood Targeting*. London: Wiley.

Hincks, S., Kingston, R., Webb, B. and Wong, C. (2018). A New Geodemographic Classification of Commuting Flows for England and Wales. *International Journal of Geographical Information Science*, 32(4), 663–684.

Johnston, R.J. (1976). *Classification in Geography*. Kent: The Invicta Press.

Jolliffe, I.T. (2002). *Principal Component Analysis*. Berlin: Springer.

Kaufman, L. and Rousseeuw, P.J. (2009). *Finding Groups in Data: An Introduction to Cluster Analysis*. Hoboken, NJ: John Wiley & Sons.

Kirch, W. (Ed) (2008). *Encyclopedia of Public Health*. Dordrecht: Springer.

Leventhal, B. (2016). *Geodemographics for Marketers: Using Location Analysis for Research and Marketing*. London: Kogan Page.

Milligan, G.W. and Stephen, C.H. (2003). Clustering and Classification Methods. In: B.W. Irvin, J.A. Schinka and W.F. Velicer (Eds), *Handbook of Psychology*. Hoboken, NJ: John Wiley & Sons.

Ojo, A. (2009). *A Proposed Quantitative Comparative Analysis for Geodemographic Classifications*. Yorkshire and Humber Public Health Observatory Technical Proceeding. York: Yorkshire and Humber Public Health Observatory.

Ojo, A. and Ezepue, P.O. (2011). How Developing Countries Can Derive Value from the Principles and Practice of Geodemographics, and Provide Fresh Solutions to Millennium Development Challenges. *Journal of Geography and Regional Planning*, 4(9), 505–512.

Ojo A. and Ojewale, O. (2019). *Urbanization and Crime in Nigeria*. Cham: Palgrave Macmillan.

Ojo, A., Papachristodoulou, N. and Ibeh, S. (2018). The Development of an Infrastructure Quality Index for Nigerian Metropolitan Areas Using Multivariate Geo-Statistical Data Fusion. *Urban Science*, 2(3), 59.

Qin, S., Kin, J., Arafat, D. and Gibson, G. (2012). Effect of Normalization on Statistical and Biological Interpretation of Gene Expression Profiles. *Frontiers in Genetics*, 3, 160.

Raykov, T. (1998). Cronbach's Alpha and Reliability of Composite with Interrelated Non-Homogenous Items. *Applied Psychological Measurement*, 22(4), 375–385.

Rummel, R.J. (1967). Understanding Factor Analysis. *Journal of Conflict Resolution*, 11(4), 444–480.

Schachter, H.L. (2010). Objective and Subjective Performance Measures: A Note on Terminology. *Administration and Society*, 42(5), 550–567.

Singleton, A.D. and Longley, P. (2015). The Internal Structure of Greater London: A Comparison of National and Regional Geodemographic Models. *Geography and Environment*, 29(1), 69–87.

Vickers, D.W. (2006). *Multi-Level Integrated Classifications Based on the 2001 Census*. PhD thesis, School of Geography, University of Leeds, United Kingdom.

Vickers, D. and Rees, P. (2007). Creating the UK National Statistics 2001 Output Area Classification. *Journal of the Royal Statistical Society: Series A (Statistics in Society)*, 170(2), 379–403.

Voas, D. and Williamson, P. (2001): The Diversity of Diversity: A Critique of Geodemographic Classification. *Area*, 33(1), 63–76.

Walker, J.T. and Maddan, S. (2008). *Statistics in Criminology and Criminal Justice: Analysis and Interpretation*. Sudbury, MA: Jones and Bartlett Publishers

Wallace, M. and Denham, C. (1996). *The ONS Classification of Local and Health Authorities of Great Britain*. Studies on Medical and Population Subjects, ONS. London: Her Majesty's Stationery Office, vol. 59.

Yiannakis, A. (1992). Toward an Applied Sociology of Sport: The Next Generation. In: A. Yiannakis and S. Greendorfer (Eds), *Applied Sociology of Sport*, Champaign, IL: Human Kinetics Publishers.

6

Machine Learning Methods for Building Small Area Classifications

6.1 Machine Learning, Artificial Intelligence, and GIS

New and emerging forms of technologies and sensors have created opportunities to generate enormous data sets often called "big data." Big data may be described as data sets that are characterized by a variety of data styles and require rapid processing speeds. In general, they are challenging to manage with traditional data analysis platforms. The characteristics of big data are often summarized with the five Vs, which denote volume, velocity, variety, veracity, and value (Hadi et al., 2015). Volume is about the size and scale of data. There is no specific agreement in the academic literature about the minimum size of data that can be described as a "big data set." Some researchers have argued that big data sets require terabytes (240 TB) or petabytes (250 PB) of storage space (The Economist, 2010). The concept of velocity refers to the speed of generating, collecting, and analyzing vast amounts of data. Variety is about the different forms of data that can be used for understanding human and natural processes. The diversity of data we now have are presented in structured and unstructured formats. Due to the rapid and voluminous generation of data, a lot of noise tends to be introduced and sometimes questions are asked about the quality of big data sets especially for human-centered research (Waldherr, 2010). Veracity focuses on the accuracy and trustworthiness of big data. Finally, value questions the notion of some that suggests that the more data there is, the better it is. Value refers to the actual worth of big data. It is not useful to have stocks of endless volumes of data if the data cannot be converted into useful solutions.

In recent decades, the exponential rise in the generation of big data sets has continued to shape the curiosity of computer scientists. This curiosity has led researchers to investigate the ability of computer programs to think and behave like humans, forming the basis of artificial intelligence (AI) and

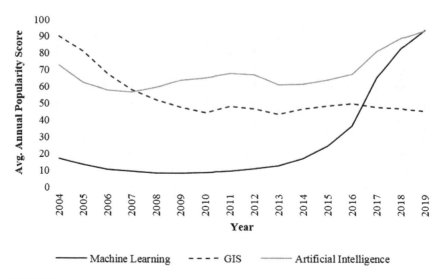

FIGURE 6.1

Google search trends for machine learning, GIS, and artificial intelligence, 2004–2019 (Author's elaboration based on data from Google Trends).

machine learning (ML). Although AI and ML are often used interchangeably, they mean different things. In a sense, AI is viewed as the umbrella term from which ML has evolved. While AI refers more generally to computer programs being able to think and behave like humans, MI goes beyond that by providing machines with the data they require to learn how to act without being explicitly programmed to do so (Joshi, 2020). MI helps to make sense of noisy data by searching for hidden patterns.

Recently, GIS software developers have been able to incorporate learning layers into their software in order to add context to geographic analysis. These ML capabilities enable GIS software to automatically discover patterns in big data sets (through classification, prediction, and segmentation) that lead to actionable insights. Figure 6.1 presents the global popularity of ML, GIS, and AI between 2004 and 2019. The numbers represent the average annual search interests on Google Search. A value of 100 is the peak popularity for the term. A value of 50 means that the term is half popular. A score of 0 means there was not enough data for this term. From the Google Trends popularity graph shown in Figure 6.1, one can observe that GIS was the more popular search term until AI surpassed it in 2007. However, ML has steadily grown in popularity, surpassing GIS in 2016 and catching up with AI in 2019. This is an indication that future GIS applications must continue to find ways to incorporate sophisticated ML capabilities.

6.2 Algorithms for Grouping Data

Although the term ML has become popular, question arises as to what exactly is the machine? "Machine" is essentially a term that denotes the statistical model or algorithm used to discover patterns in big data that lead to actionable insights. There are different types of algorithms, and based on the way they learn about patterns in data, they can be broadly classified into supervised and unsupervised learning algorithms.

A supervised learning algorithm is used to train model on known input and output data so that it can predict future outputs (Joshi, 2020). Supervised learning is so named because the researcher deploying the algorithm acts as a guide to teach the algorithm what conclusions it should reach. Essentially, supervised learning requires that the algorithm's possible outputs are already known and that the data used to train the algorithm are already labeled with correct answers as illustrated in Figure 6.2.

Unsupervised learning algorithms are more closely aligned with true AI because of their capability to learn to identify complex processes and patterns without human intervention (Joshi, 2020). While supervised learning algorithms require labeled inputs, unsupervised learning algorithms can draw references from data sets consisting of input data without labeled responses as illustrated in Figure 6.3.

Unsupervised learning algorithms are generally preferred for building small area geodemographic classifications because they can be used to find meaningful structure, group data, and reveal explanatory underlying processes. There are several examples of unsupervised learning algorithms, which cannot be completely exhausted in this chapter. However, to facilitate the discussion of some of the key algorithms, they are categorized under four clustering methods, which include hierarchical, partitional, density-based, and grid-based clustering methods.

6.2.1 Hierarchical Clustering Methods

Hierarchical clustering methods are used to group data about people and places by ordering them in the form of a pyramid. In other words, these clustering methods construct trees of clusters of objects such that any two groups are disjoint, or one includes the other (Jain et al., 1999). There are two classes of hierarchical clustering methods: agglomerative and divisive hierarchical clustering.

6.2.1.1 *Agglomerative Hierarchical Clustering*

This is the more common hierarchical clustering approach. It is a bottom-up clustering approach where each individual clustering object (e.g. small

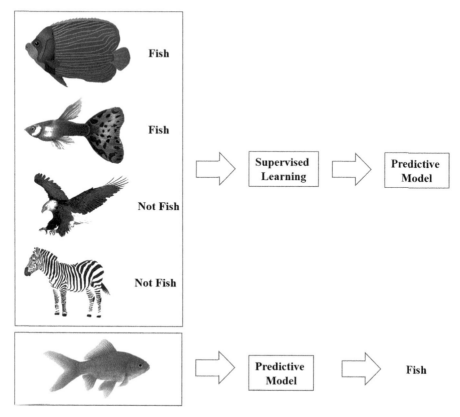

FIGURE 6.2
An illustration of supervised learning.

geographic areas) starts in its own cluster, and pairs of clusters are merged iteratively as one moves up the hierarchy (Day and Edelsbrunner, 1984). The algorithm works by initially computing a distance matrix between different objects. The two closed objects are then merged based on a distance measure or criterion. There are different ways to measure the distance between clusters and they are often called linkage methods. The *complete linkage* distance measure is defined as the longest distance between two points in each cluster. The *single-linkage* distance measure is defined as the shortest distance between two points in each cluster. The *average linkage* distance measure is defined as the average distance between each point in one cluster and every point in the other cluster. Finally, the *centroid linkage* distance measure finds the centroid of cluster 1 and centroid of cluster 2 and then calculates the distance between the two before merging. Once two objects are merged based on a chosen linkage measure, the distance matrix is updated and the process repeats itself until one single cluster remains. Figure 6.4 shows an illustration of the agglomerative clustering procedure.

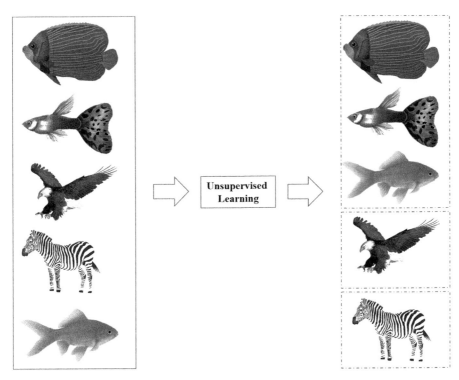

FIGURE 6.3
An illustration of unsupervised learning.

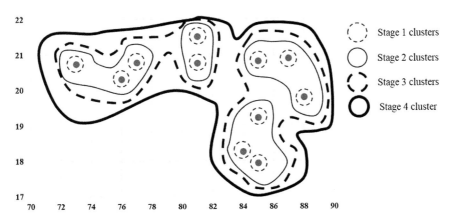

FIGURE 6.4
An illustration of agglomerative hierarchical clustering.

Joe H. Ward developed the Ward's hierarchical clustering algorithm in 1963 during his service at the Aerospace Medical Division of Lackland Air Force Base. Ward's hierarchical clustering approach developed in 1963is arguably the most commonly used agglomerative hierarchical data grouping method, although some scholars have questioned its efficiency. Everitt et al. (2001) reported that the algorithm has the tendency to impose artificial spherical clusters.

$$\Delta(A,B) = \sum_{i \in A \cup B} \left\| \vec{x}_i - \vec{m}_{A \cup B} \right\|^2 - \sum_{i \in A} \left\| \vec{x}_i - \vec{m}_A \right\|^2 - \sum_{i \in B} \left\| \vec{x}_i - \vec{m}_B \right\|^2 \tag{6.1}$$

$$= \frac{n_A n_B}{n_A + n_B} \left\| \vec{m}_A - \vec{m}_B \right\|^2 \tag{6.2}$$

where:

$\vec{m}_j$ represents the center of cluster j
$\vec{n}_j$ represents the number of points in the center of cluster j
Δ represents the merging cost of combining clusters A and B.

Ward (1963) suggested that the distance between two clusters is based on how much the sum of squares will increase when they are merged, as expressed in Equations 6.1 and 6.2. Since every point is a cluster at the beginning of the process, the sum of squares starts out at zero and increases slowly as clusters are merged. Romesburg (2004) observed that a disadvantage of this algorithm is that the hierarchy imposes a restriction on the clustering objects, making it impossible for objects to move between clusters.

6.2.1.2 Divisive Hierarchical Clustering

Divisive hierarchical clustering works in a manner opposite to the agglomerative algorithm. Divisive algorithms employ a top-down approach. All clustering objects are initially assigned to a single cluster. The single cluster is then partitioned into two and the partitioning process continues until all objects eventually form their own clusters. Compared with agglomerative approach, the divisive approach is less widely used due to its complexity although evidence suggests that divisive algorithms produce more accurate hierarchies than agglomerative algorithms in certain circumstances (Bouguettaya et al., 2015).

An example of divisive hierarchical clustering algorithm that is growing in terms of its use is a method named clustering using representatives (CURE). The steps for implementing CURE are summarized in Figure 6.5 according to Guha et al. (1998). One of the general drawbacks of hierarchical clustering algorithms is that they struggle with outliers and they have a tendency to generate spherical clusters. However, it has been suggested that CURE is a more efficient hierarchical data grouping method that has the advantage of

FIGURE 6.5
Steps for implementing CURE algorithm.

coping with outliers and detecting nonspherical shapes (Guha et al., 1998). Despite its identified merits, CURE cannot efficiently handle different densities and the estimation of its complexity is not straightforward. Furthermore, its complexity and high runtime mean that it struggles with large databases.

Other useful hierarchical clustering methods include CHAMELEON, which uses a graph partitioning method (Karypis, 1999); COBWEB, which works by updating clusters object by object (Fisher, 1987); and Self-Organizing Maps, which function similar to artificial neural networks (Kohonen, 1990).

6.2.2 Partitional Clustering Methods

While hierarchical clustering methods generate a set of nested clusters that are organized in the form of a tree, partitional methods work differently. Partitional clustering divides a set of data objects into nonoverlapping subsets such that each data object is in exactly one subset. Generally, these methods are used to decompose a data set into a set of disjointed groups. Unlike hierarchical approaches, which can commence with any clustering seed, partitional clustering methods need to be provided with a set of initial clusters as a starting point.

One of the most widely used partitional clustering methods is the *k*-means algorithm. *K*-means clustering is an unsupervised ML algorithm, which is used when data are without defined categories. The goal of this algorithm is to search for groups in the data, with the number of groups represented by the variable *K*. The algorithm works iteratively to assign each clustering object to one of *K* groups based on the variables that are provided. Objects are clustered based on a measure of variable similarity (Harris et al., 2005).

The objective of *k*-means clustering is to minimize the sum of squared distances between all points and the cluster center. The algorithm first creates groups of similar objects by imperfectly arranging them into the final number of predefined clusters and rearranging the assignments iteratively in order to generate a better fit as denoted in Equation 6.3. This method of cluster reassignment helps to reduce the variability within groups and increase the variability between them.

$$V = \sum_{k}^{y-1} \sum_{v}^{x=1} \sum_{n_k}^{i=1} \left(z_{yxi} - \mu_{xy} \right) \tag{6.3}$$

TABLE 6.1
Variants of the *K*-Means algorithm

Type of Modification Carried Out on *K*-Means	Algorithms
Extended to select different representative prototypes, and for improving clusters' resilience to outliers (Mirkin, 2005). Extended to handle non-numerical attributes and categorical data (Xu and Wunsch, 2005).	• *K*-medoids • *K*-medians • *K*-modes
Extended to select better initial centroid estimates and tackle the problem of converging to a local minimum (Krishna and Murty, 1999; Mirkin, 2005).	• Intelligent *K*-means • Genetic *k*-means
Extended by applying feature weighting techniques (Hartigan and Wong, 1979; Scholkopf et al., 1998; Huang et al., 2005, p. 44).	• Weighted *k*-means • Kernel *k*-means

where:

V represents the sum of squared distances of all variables from cluster means for all clusters

z_{yxi} represents the standardized variable for object i, variable x, and cluster y

μ_{xy} represents the mean for variable x in cluster y

k represents the number of clusters

v represents the number of variables

n_k represents the number of objects in a cluster.

Due to its flexibility, it is possible to adapt and build upon the *k*-means algorithm in order to develop algorithms that are more efficient. In Table 6.1, other prominent variants of the *k*-means algorithm are presented together with the types of modifications carried out on them.

6.2.3 Density-Based Clustering Methods

An underlying assumption of many well-known clustering algorithms is that grouped data are generated from probability distributions. This assumption is the major reason as to why most algorithms generate spherical clusters. In many spatial data sets, nonspherical clusters tend to occur naturally, and it has been suggested that hierarchical and partitional methods struggle with grouping such data sets (Vickers, 2006). Furthermore, real-life spatial data sets that require clustering of areal units are increasingly large and presenting additional challenges of arbitrariness and scalability.

The competing challenges confronting many hierarchical and partitional clustering methods have motivated the development of density-based clustering approaches. Density-based methods have the capability to identify and remove outliers and noise that exist in large spatial data sets

(Shrivastava and Gupta, 2012). Another strength of these methods is that they embrace nonparametric approaches, which desist from making prior assumptions about the probable number of clusters existing in a distribution or their possible patterns in taxonomic space. According to Saraswathi and Sheela (2014), density-based methods work by considering clusters as dense regions of objects in the data space that are separated by regions of low density representing noise. Density-based clustering algorithms should be able to answer some fundamental questions. They should be able to clarify how density is estimated. They should elucidate how the connectivity of clustered objects is defined. Finally, they should expose those data structures that underpin the efficient application of the algorithm.

One of the most popular density-based clustering methods is density-based spatial clustering of applications with noise (DBSCAN). When deployed on a spatial data set, the algorithm searches for areas of highly concentrated data points and highlights those groups that are suitably dense – as defined by the input parameters (Ester et al., 1996). At the end of the algorithm, every object will have been assigned to a cluster or identified as noise.

DBSCAN approximates density by counting the number of objects in a fixed-radius neighborhood. The radius defines the area of interest (shape and size) around each object and the minimum density threshold sets the minimum number of points, which must fall within this area in order for it to be considered dense. Two objects are considered to be connected if they lie within each other's neighborhood.

Furthermore, during the process of determining those objects that can belong to the same cluster, DBSCAN assigns objects as core, boundary, and noise points. A point is said to be a core point if within the area of interest there are a greater number of other data points than the minimum point's threshold. Boundary points do not meet the minimum density threshold but have at least one core point within their areas of influence. Noise points do not meet the minimum density threshold and do not fall sufficiently close to a core point. Clusters are built by grouping core points that fall within each other's areas of influence.

Figure 6.6 is a graphical representation of how the DBSCAN algorithm works. In this example, the minimum points threshold is four. The ovals drawn with dashed lines represent the areas of influence of the four points to which arrows are pointing. Each of the core points in this illustration have four or more other points within their areas of influence (i.e. including the core points). The boundary points do not reach the minimum points threshold of four. However, they subsume at least one core point within their areas of influence. The noise points do not satisfy any of the tests (i.e. they do not reach the minimum points threshold and do not have a core point within their areas of influence. The thick black polygon shows the convex outline of the cluster.

Other useful density-based clustering algorithms include density-based clustering (DENCLUE), which is considered as a special case of the kernel density estimation (KDE) (Hinneburg and Keim, 1998), and ordering points to identify clustering structure (OPTICS) (Ankerst et al., 1999).

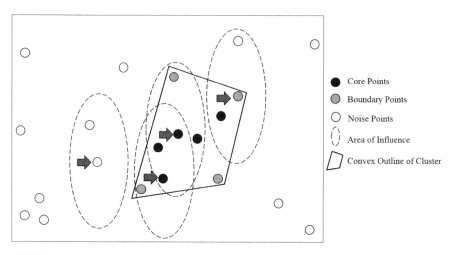

FIGURE 6.6
An illustration of how DBSCAN algorithm works.

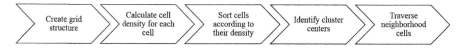

FIGURE 6.7
General steps for implementing grid-based algorithms.

6.2.4 Grid-Based Clustering Methods

In grid-based clustering, the data space is divided into a finite number of cells that form a grid-like structure. The clustering algorithm is then applied on the grid, rather than directly on the database. The emerging clusters are characterized by greater density of data points than their surroundings. Grid-based clustering has the advantage of reducing time complexity for very large data sets. The reduction in time complexity is achieved by the way the algorithms work. Since the data points in a data set are converted to a grid, one ends up with a significantly smaller number of cells between data points. The grid-based approach clusters the smaller number of cells (i.e., the neighborhoods surrounding the data points), thereby greatly improving its speed and efficiency. It has been suggested that grid-based algorithms work through five steps (Grabusts and Borisov, 2002; Ma et al., 2010). These steps are illustrated in Figure 6.7.

The statistical information grid-based clustering method (STING) is one of the classical examples of grid-based techniques for clustering spatial data sets. The STING algorithm splits the entire geographic area into rectangular cells and stores the cells in a hierarchical grid structure (Wang et al., 1997) as

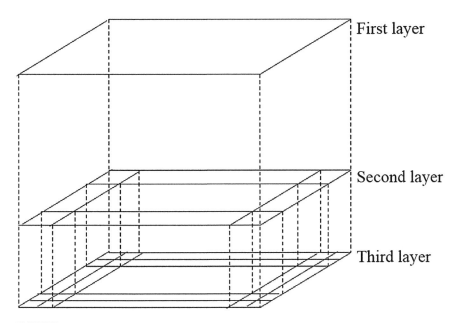

FIGURE 6.8
Hierarchical grid structure tree.

illustrated in Figure 6.8. The different layers in the cell structure correspond to different levels of resolution.

As shown in Figure 6.8, each cell at a high-level layer is partitioned into a number of smaller cells in the next lower level. Statistical information of each cell is calculated and stored beforehand and is used to answer queries. Additionally, the parameters of higher-level cells can be easily estimated from the parameters of lower- level cells.

The STING algorithm has been further extended to STING+ to enable it to process dynamically evolving spatial databases and to enable active datamining (Wang et al., 1999). Other grid-based methods used for clustering large data sets include grid-based hierarchical clustering (GRIDCLUS) (Schikuta, 1996), WaveCluster (Sheikholeslami et al., 1998), Adaptive Mesh Refinement (Liao et al., 2004), and clustering in quest (CLIQUE) (Agrawal et al., 1998).

6.3 Determining the Ideal Number of Groups

What is the ideal number of clusters? This is one of the main questions encountered during the development of small area classifications.

Researchers often grapple with how to determine an appropriate number of cluster groups after deploying certain types of ML algorithms such as the partitional methods. In countries where previous classifications have been built and tested, there are guiding frameworks that can be used. However, in developing countries where these types of area classifications have not been built, researchers grapple with nebulous ideas about the probable number of representative groups that may be inherent within their data. This is because there are no previous examples against which to compare new area classification outputs.

It is broadly accepted that the methods used to investigate the natural number of clusters present within large spatial data sets are slightly informal. Consequently, a wide range of strategies for estimating the optimal number of clusters have been proposed. Everitt et al. (2001) suggested that after creating several solutions with varying numbers of clusters, the value of a clustering criterion for each solution could be plotted against the number of clusters created, and the point where one observes an abrupt change in the plotted graph, as illustrated in Figure 6.9, signifies the representative number of clusters. This method is also called the "elbow" method. In the example shown, the first major point of sudden change (elbow) in the clustering criterion is at the solution for 11 clusters.

Miligan and Cooper (1985) conducted an extensive comparative evaluation of 30 possible methods for estimating the true number of clusters when applying hierarchical ML algorithms to simulated data with well-separated clusters. According to their results, Calinski and Harabasz's index (1974) is the most effective measure, followed by Duda and Hart's method (1973) and the C-index (Dalrymple-Alford, 1970). These three measures are summarized in Table 6.2. Islam et al. (2015) have discussed several other measures for detecting the ideal number of clusters after deploying an ML algorithm.

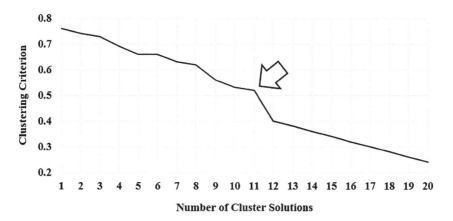

FIGURE 6.9
Using the elbow method to detect the probable number of groups.

TABLE 6.2
Prominent measures for determining the number of clusters

Criterion	Mathematical Notation	Comments
Calinski–Harabasz (CH) index	$CH_k = \dfrac{SSB_k / (K-1)}{SSW_k / (n-K)}$ where: K represents the number of clusters; n represents the number of data points; SSB_k and SSW_k represent between- and within-cluster sum of squares, respectively.	The number of clusters is chosen by maximizing the function given in the equation.
Duda–Hart (DH) index	$DH_K = \dfrac{SSW_{K+1}}{SSW_K}, K \geq 1$ $DH_K < \left[1 - \dfrac{2}{\pi p} - \Delta \sqrt{\dfrac{2\left\{1 - 8\left(\pi^2 p\right)\right\}}{n_c p}} \right]$ where: K represents the number of clusters; z represents a standard normal quantile (suggested value is 3.20); n_c represents the number of observations of cluster c that was split up to form $K+1$ clusters in total.	The first equation generates the Duda– index. It is essentially the ratio of within sum of squares. If the index is less than the criterion shown in the second equation, then a cluster will be subdivided.
C-index	$C_k = \dfrac{S_{wk} - S_{min}}{S_{max} - S_{min}}$ where: K represents the number of clusters; S_{wk} represents the sum of distances between all the pairs of points inside each cluster; S_{min} represents the sum of the smallest distances between all the pairs of points in the entire data set; S_{max} represents the sum of the largest distances between all the pairs of points in the entire data set.	The C-index ranges between 0 and 1. The ideal number of clusters will be the number that minimizes the index.

Sometimes, evaluating the clustering criterion against the number of clusters may not clearly reveal the probable number of clusters. Other rules of thumb that have been recommended by Vickers (2006) in his seminal work include:

- Prioritizing the larger number of clusters if it is difficult to choose between two solutions
- Selecting the cluster that shows the greatest reduction in the average distance from the solution with one having fewer clusters in a nonhierarchical system

- Selecting the solution that shows the greatest increase in the average distance between the most dissimilar objects within merged clusters in a hierarchical system
- Selecting the solution that has the most suitable number of clusters for purpose
- Selecting the solution that is most homogeneous in terms of the number of objects within each cluster.

In a nutshell, there is no one-size-fits-all solution for dealing with the problem of estimating the correct number of groups after clustering small area data sets.

6.4 Sensitivity Analysis

All forms of geographic data are exposed to some form of uncertainty (Unwin, 1995). Uncertainty simply means that due to certain factors (limitation of human knowledge, digital re-presentation of real-world data, measurement errors, etc.) there may be some doubt about spatial data and analysis. The uncertainty that is inherent within geographic data also means that further discrepancies and distortions can be introduced during analysis of the spatial data. Several assumptions are also made in the course of developing small area classifications (Harris, 2005). These underlying assumptions can also introduce some errors in the analysis. This is why sensitivity analysis is considered a necessary requirement during the analysis of big multivariate spatial data sets.

It is also important to recognize that during the preprocessing of data, variable selection, normalization, standardization, weighting, and choice of ML algorithm may inadvertently favor some preconception of the real world and weaken others. These competing activities serve as additional stimulus for sensitivity testing and verification.

Sensitivity analysis allows us to study and quantify the variations in the uncertainty introduced into the outputs of analysis (Crosetto et al., 2000). It also yields quantitative and qualitative insights into how the outputs might change under different scenarios. Sensitivity analysis is also known as "what-if" analysis. In general, a key reason for performing sensitivity analysis during the development of a small area geodemographic classification is to test how each independent variable will affect the composition of clusters generated from the analysis under a given set of assumptions.

Within the context of small area geodemographic classifications, sensitivity analysis works on a simple principle: change the inputs and observe the outputs. In order to test for the effect of input variables on the classification, informal sensitivity analysis based on alternative cluster solutions can be carried out. This involves conducting repeated clustering reruns such that an input variable is excluded during each rerun. For instance, if 50 input variables are used to produce an area classification system, the ML algorithm can be rerun 50 times on 49 variables per time to test the influence of each variable on the classification. After the 50 reruns, the average distance of each clustered object from its cluster center can be compared with each other to see whether there is a significant difference. Vickers (2006) found that by evaluating the difference that reclustering $n-1$ variables makes to the average difference from the cluster center, a judgement can be made about the effect of each variable on the classification.

6.4.1 Tukey's Post Hoc Test and Analysis of Variance (ANOVA)

Cluster distances can also be evaluated using a diagnostic statistic such as Tukey's post hoc test.

This test is a statistical tool used to determine whether the relationship between two sets of data is statistically significant (Crawshaw and Chambers, 2001). It can indicate whether there is a strong chance that an observed numerical change in one value is causally related to an observed change in another value.

$$\text{HSD} = \frac{M_i - M_j}{\sqrt{\dfrac{MS_w}{n}}} \tag{6.4}$$

where:

M_i represents the larger of the two means being compared
M_j represents the smaller of the two means being compared
MS_w represents the within-group variance
n represents the size of the sample.

Given in Equation 6.4, Tukey's test can be invoked when there is a need to determine whether the distances between cluster centroids are statistically significant. In conjunction with the Tukey's test, it is recommended that one-way ANOVA should be calculated at each step of the algorithm rerun.

6.4.2 Test of Discriminatory Power

Another pseudo-sensitivity test that could be carried out after repeated reclustering of $n-1$ variables is the test of discriminatory power. The discriminatory power of an area classification tells us how effective the classification is in distinguishing and segmenting different population characteristics by area typologies. The use of Gini coefficient and Lorenz curve has been suggested as useful approaches for assessing the discriminatory power of small area classifications (Leventhal, 1995).

After repeated reclustering of $n-1$ variables, the Gini coefficient denoted in Equation 6.5 can be calculated after each rerun to detect how the exclusion of each of the input variables influences the overall discriminatory power. The coefficient allows a graphical comparison of inequality using a Lorenz curve. Values for the Gini coefficient range from 0, where there is perfect equality, to 1, where there is perfect inequality. The Gini coefficient represents an expression of the area located between the line of perfect equality and the Lorenz curve. Some researchers have questioned the use of the Gini coefficient and the Lorenz curve. However, Leventhal (1995) described it as a method that can help mitigate the challenges posed by numerical methods of evaluation. Specifically, he suggested that numerical methods are unable to assess the usefulness of a discriminator and may not be able to control for the different number of clusters characteristic of different classifications.

$$G = 1 - \sum_{i=0}^{k=1} \left(y_{i+1} + y_i \right) \left(x_{i+1} - x_i \right) \tag{6.5}$$

where:

- G represents the value of the Gini coefficient
- k represents the number of data points for the observed variable and population base
- y represents the observed variable for a selected area classification cluster
- x represents the population base for the corresponding area classification cluster.

To calculate the Gini coefficient for each area classification, external variables (such as population of interest and a corresponding base population) are required (Ojo, 2009). For instance, the population of interest might be the *percentage of adults who voted in an election*. The corresponding population base would be *all adults who are eligible to vote*. A measure such as rate could be used to sort the population of interest by their cluster divisions. Following this, the percentage shares of the population of interest and population base can be accumulated and used to derive the Gini coefficient.

6.4.3 Test of Predictive Power

In addition to testing for discriminatory power, variables may also be assessed in terms of the extent to which their presence or absence affects the predictive power of an area classification. To assess the influence of variables on predictive capacity of classifications, observed and expected frequencies of external variables (not included in the clustering process) are required. The observed frequency is the actual frequency that is obtained from the field data. The expected frequency is the theoretically predicted frequency, which can be calculated by using the probability theory.

$$E^{RMS} = \left[\frac{1}{m} \sum_m \left(\frac{y - y'}{y} \right) \right]^{1/2}$$
(6.6)

where:

E^{RMS} represents the root mean square error

m represents the number of geographic areas

y represents the observed frequency for a selected variable

y' represents the expected frequency for a selected variable.

After reproducing the area classifications repeatedly using $n-1$ variables, the observed and expected frequencies of the external variables should be aggregated by cluster divisions. The difference between observed and expected frequencies can be taken and the error introduced can be quantified by using a variant of Fisher and Langford's (1995) root mean square (RMS) error adapted by Gregory (2000) (see Equation 6.6). The lower the value of the RMS, the better the predictive power of the classification.

6.5 Naming and Describing Groups

Labeling a group with a name can be contentious. It is a complex process requiring consideration of numerous issues. The names assigned to different clusters should be representative of the broad characteristics of population linked to them. The names attached to clusters are only indicative of the predominant features of the areas in question. Some guiding principles for labeling clusters include:

- Comparing the distribution of objects in one cluster with that of other clusters
- Evaluating the geographic distribution and concentration of members of different clusters
- Examining between- and within-cluster variations

- Determining the cluster centroid
- Using a list of variables (terms) with high weights in the centroid of the cluster.

Names should not in any way be offensive especially in multiethnic or multireligious countries. Care should be taken to ensure that the chosen labels do not stigmatize any section of the population.

Cluster profiles and pen portraits are textual and graphical descriptions that summarize the prevalent characteristics of each cluster. They have the benefit of elucidating (in qualitative terms) some of the information inherent in complex quantitative analysis. Cluster profiles and pen portraits can also be developed for qualitative assimilation of the key characteristics of clusters.

6.6 Conclusion

AI is an umbrella term used for systems that demonstrate at least some of the following behaviors associated with human intelligence: planning, learning, reasoning, problem-solving, knowledge representation, perception, motion, and manipulation, and, to a lesser extent, social intelligence and creativity. One of the subfields of AI is ML, which has grown rapidly in recent decades partly due to an explosion of large-scale data generation. ML makes sense of noisy data; therefore, it is useful for addressing spatial pattern recognition and classification problems. The strengths and weaknesses of multiple ML approaches for classifying geographic data have been explored in this chapter. Some algorithms are incapable of automatically detecting the number of natural clusters within large data sets. Therefore, some of the methods for determining the ideal number of clusters have also been explored. There is no one-size-fits-all solution. Sensitivity analysis is important for addressing the pitfalls of uncertainty and assumptions introduced during data preprocessing activities. The labeling of clusters is both an art and a science. The chapter concludes by providing some good-practice considerations that should be borne in mind when labeling clusters.

References

Agrawal, R., Gehrke, J., Gunopulos, D. and Raghavan, P. (1998). Automatic Subspace Clustering of High Dimensional Data for Data Mining Applications. *Proceedings* of the 1998 ACM SIGMOD International Conference on Management of Data, SIGMOD, 94–105.

Ankerst, M., Breunig, M.M., Kriegel, H.-P. and Sander, J. (1999). OPTICS: Ordering Points to Identify the Clustering Structure. *Special Interest Group on Management of Data (SIGMOD) Conference*, Seattle, Washington, 2–4 June, 49–60.

Bouguettaya, A., Yu, Q., Liu, X., Zhou, X. and Song, A. (2015). Efficient Agglomerative Hierarchical Clustering. *Expert Systems with Applications*, 42(5), 2785–2797.

Calinski, T. and Harabasz, J. (1974). A Dendrite Method for Cluster Analysis. *Communications in Statistics*, 3(1), 1–27.

Crawshaw, J. and Chambers, J. (2001). *A Concise Course in Advanced Level Statistics with Worked Examples*. Cheltenham: Nelson Thornes.

Crosetto, M., Tarantola, S. and Saltelli, A. (2000). Sensitivity and Uncertainty Analysis in Spatial Modelling Based on GIS. *Agriculture, Ecosystems and Environment*, 81(1), 71–79.

Dalrymple-Alford, E. (1970). Measurement of Clustering in Free Recall. *Psychological Bulletin*, 74(1), 32–34.

Day, W.H.E. and Edelsbrunner, H. (1984). Efficient Algorithms for Agglomerative Hierarchical Clustering Methods. *Journal of Classification*, 1(1), 7–24.

Duda, R.O. and Hart, P.E. (1973). *Pattern Classification and Scene Analysis*. New York, NY: Wiley.

Ester, M., Kriegel, H.-P., Sander, J. and Xu, X. (1996). A Density-Based Algorithm for Discovering Clusters in Large Spatial Databases with Noise. *Proceedings of International Conference on Knowledge Discovery and Data Mining*, 226–231.

Everitt, B.S., Landau, S. and Leese, M. (2001). *Cluster Analysis*. London: Arnold.

Fisher, D.H. (1987). Knowledge Acquisition via Incremental Conceptual Clustering. *Machine Learning*, 2(2), 139–172.

Fisher, P.F. and Langford, M. (1995). Modeling the Errors in Areal Interpolation between Zonal Systems by Monte Carlo Simulation. *Environment and Planning A*, 27(2), 211–224.

Gregory, I.N. (2000). *An Evaluation of the Accuracy of the Areal Interpolation of Data for the Analysis of Long-Term Change in England and Wales*. Available at: www.geocomputation.org/2000/GC045/Gc045.htm.

Guha, S., Rastogi, R. and Shim, K. (1998). CURE: An Efficient Clustering Algorithm for Large Databases. *Information Systems*, 26(1), 35–58.

Hadi, H.J., Shanain, A.H., Hadishaheed, S. and Ahmad, A.H. (2015). Big Data and the Five V's Characteristics. *International Journal of Advances in Electronics and Computer Science*, 2(1), 16–23.

Harris, R., Sleight, P. and Webber, R. (2005). *Geodemographics, GIS and Neighborhood Targeting*. London: Wiley.

Hartigan, J.A. and Wong, M.A. (1979). Algorithm as 136: A K-means Clustering Algorithm. *Journal of the Royal Statistical Society. Series C*, 28(1), 100–108.

Hinneburg, A. and Keim, D.A. (1998). An Efficient Approach to Clustering in Large Multimedia Databases with Noise. *Proceedings of the 4th International Conference on Knowledge Discovery and Data Mining*, New York, NY, 98, 58–65.

Huang, J.Z., Ng, M.K., Rong, H. and Li, Z. (2005). Automated Variable Weighting in K-Means Type Clustering. *IEEE Transactions on Pattern Analysis and Machine Intelligence*, 27(5), 657–668.

Islam, M.A., Alizadeh, B.Z., van den Heuvel, E.R., Bruggeman, R., Cahn, W., de Haan, L., Kahn, R.S., Meijer, C., Myin-Germeys, I., van Os, J. and Wiersma, D. (2015).

A Comparison of Indices for Identifying the Number of Clusters in Hierarchical Clustering: A Study on Cognition in Schizophrenia Patients. *Communications in Statistics: Case Studies, Data Analysis and Applications*, 1(2), 98–113.

Jain, A.K., Murty, M.N. and Flynn, P.J. (1999). Data Clustering: A Review. *ACM Computing Surveys (CSUR)*, 31(3), 264–323.

Joshi, A.V. (2020). *Machine Learning and Artificial Intelligence*. Cham: Springer Nature.

Karypis, G., Han, E.H. and Kumar, V. (1999). CHAMELEON: Hierarchical Clustering Using Dynamic Modeling. *Computer*, 32(8), 68–75.

Kohonen, T. (1990). The Self-organizing Map. *Proceedings of the IEEE*, 78(9), 1464–1480.

Krishna, K. and Murty, M.N. (1999). Genetic K-Means Algorithm. *IEEE Transactions on Systems, Man, and Cybernetics, Part B: Cybernetics*, 29(3), 433–439.

Leventhal, B. (1995). Evaluation of Geodemographic Classifications. *Journal of Targeting, Measurement and Analysis for Marketing*, 4(2), 173–183.

Liao, W.-K., Liu, Y. and Choudhary, A. (2004). A Grid-Based Clustering Algorithm Using Adaptive Mesh Refinement. *Proceedings of the 7th Workshop Mining Scientific and Engineering Datasets*, Lake Buena Vista, Florida, 24 April, 61–69.

Milligan, G.W. and Cooper, M.C. (1985). An Examination of Procedures for Determining the Number of Clusters in a Dataset. *Psychometrica*, 50(2), 159–179.

Mirkin, B.G. (2005). *Clustering for Data Mining: A Data Recovery Approach*. Boca Raton, FL: CRC Press.

Ojo, A. (2009). *A Proposed Quantitative Comparative Analysis for Geodemographic Classifications*. York: Yorkshire and Humber Public Health Observatory.

Romesburg, C. (2004). *Cluster Analysis for Researchers*. NC, Morrisville: Lulu Press.

Saraswathi, S. and Sheela, M.I. (2014). A Comparative Study of Various Clustering Algorithms in Data Mining. *International Journal of Computer Science and Mobile Computing*, 3(11), 422–428.Schikuta, E. (1996). Grid-Clustering: An Efficient Hierarchical Clustering Method for Very Large Data Sets. *Proceedings of the 13th International Conference on Pattern Recognition*, 2, 101–105.

Scholkopf, B., Smola, A. and Muller, A.R. (1998). Nonlinear Component Analysis as a Kernel Eigenvalue Problem. *Neural Computation*, 10(5), 1299–1319.

Sheikholeslami, G., Chatterjee, S. and Zhang, A. (1998). WaveCluster: A Multiresolution Clustering Approach for Very Large Spatial Databases. *Proceedings of the 24rd International Conference on Very Large Data Bases*, 428–439.

Shrivastava, P. and Gupta, H. (2012). A Review of Density-Based Clustering in Spatial Data. *International Journal of Advanced Computer Research*, 2(3), 200–202.

The Economist (2010). *All Too Much: Monstrous Amounts of Data*. 25 February. Available at: www.economist.com/node/15557421.

Unwin, D.J. (1995). Geographical Information Systems and the Problem of Error and Uncertainty. *Progress in Human Geography*, 19(4), 549–558.

Vickers, D.W. (2006). *Multi-Level Integrated Classifications Based on the 2001 Census*. PhD thesis, School of Geography, University of Leeds, United Kingdom.

Waldherr, A., Maier, D., Miltner, P. and Günther, E. (2017). Big Data, Big Noise: The Challenge of Finding Issue Networks on the Web. *Social Science Computer Review*, 35(4), 427–443.

Wang, W., Yang, J. and Muntz. R.R. (1997). STING: A Statistical Information Grid Approach to Spatial Data Mining. *Proceedings of 23rd International Conference on Very Large Data Bases*, 186–195.

Wang, W., Yang, J. and Muntz. R.R. (1999) STING+: An Approach to Active Spatial Data Mining. *Proceedings of the 15th International Conference on Data Engineering*, 116–125.

Ward, J.H. (1963). Hierarchical Grouping to Optimize an Objective Function. *Journal of the American Statistical Association*, 58, 236–244.

Xu, R. and Wunsch, D. (2005). Survey of Clustering Algorithms. *IEEE Transactions on Neural Networks*, 16(3), 645–678.

7

Visualizing Small Area Geodemographic Data and Information Products

7.1 Goals of Visualization

Small area geodemographic data and information products are digital solutions that are structured in specific formats to instruct, educate, or guide users in order to meet predefined purposes. In developing small area classifications, the interrogation of large spatial data sets often generates multidimensional outputs (Leventhal, 2016; Webber and Burrows, 2018). Outputs are multidimensional because the development of area classifications involves a combination of complex decisions, processes, and procedures. Many of these techniques and judgments have been covered in Chapters 5 and 6 of this book. The complexity of these activities yields a remarkable mix of information products that are defined by different data and information types as illustrated in Figure 7.1. The multidimensional quality of small area geodemographic information products underscores the importance of visualization.

Categorical data and information products are characterized by discrete values, which belong to specific finite set of classes, such as the clusters to which object belong. Quantitative data and information products are numerical or measurable outputs, which may be used to produce graphs and charts. Ordinal data and information products subsume attributes that are useful in giving order to numerical data. For instance, after classification, the strength of each variable within a cluster group can be ordered to illustrate extreme, average, and below average influence. Textual data and information products accompany descriptive cluster profiles. Relational data and information products are generated in the form of trees and networks especially if hierarchical clustering algorithms are used. Since small area classification is about clustering geographies, all products often have a locational or spatial component.

Within this deluge of output types lie a wealth of valuable insights. Visualization is used as a tool for excavating insights from various forms of data and information products. Some people think about visualization as a way of bypassing tabular data and representing data in a graphical

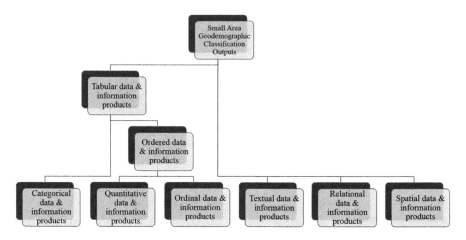

FIGURE 7.1
Multidimensional quality of small area geodemographic classification outputs.

form. However, visualization transcends the mere representation of tabular data in graphical form. It is fundamentally a mechanism for revealing and discerning the information that lies behind data using an effective method of graphical display that also helps the readership audience or viewers to see the structure within data (Chen et al., 2008).

Within the milieu of geodemographic classifications, the goals of visualization are manifold. The primary aim is to make data-driven products more accessible and more appealing to diverse audiences with different levels of statistical and technical proficiency. This can help ensure that engagement with area classification products is broad and sustained. Some of the main goals of visualization are captured in Figure 7.2. While the goals presented in Figure 7.2 are broad, they are by no means exhaustive. This is because visualization tools and techniques continue to evolve (Rom, 2015), thereby affording area classification developers the flexibility to imagine and desire new visualization objectives.

The statistical outputs that originate from complex machine learning procedures can often prove challenging for some to grasp or interpret. Visualization can be used as a tool for improving the comprehension of non-technical users of small area classification systems by helping them deduce the meanings of statistical values.

Although a prime goal for developing area classifications is to simplify complexity inherent in raw spatial data (Vickers, 2006), the procedures often end up generating too much information. Some users find the deluge of information overwhelming, while others may have too little time to comprehend too much information. Visualization methods can be combined with further spatial data analytics methods, such as exploratory spatial data analysis to yield interactive visual data analysis solutions, which help to make sense of the area classification outputs.

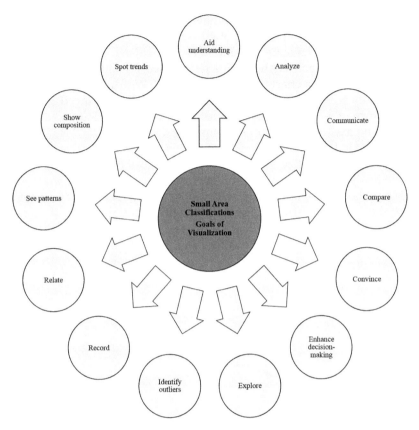

FIGURE 7.2
Visualization goals of small area geodemographic classification products.

Poorly chosen spatial visualization techniques can introduce uncertainty when communicating geodemographic information; therefore, it is important to choose visualization methods carefully. Although there are multiple visualization techniques that are being used by vendors of geodemographic systems, the remainder of this chapter explores six popular information visualization techniques within the discipline. These include radial plots, node-link trees, bar and column plots, area cartograms, and online and off-line interactive visualizations.

7.2 Radial Plots

Radial plots are one of the more popular techniques for visualizing outputs of geodemographic analysis. Galbraith (1994) describes radial plots as graphical displays for comparing estimates that have differing precisions. Geodemographic classifications can often subsume an extensive number of variables. Radial plots have been found to be particularly useful for ranking large numbers of variables. Alternative terms used for radial plots include "radar charts," "spider charts," or "web charts." With radial plots multivariate outputs from geodemographic analysis can be displayed graphically in the form of a two-dimensional chart of several quantitative variables represented on axes starting from the same point.

As illustrated in Figure 7.3, within the field of geodemographics, radial plots are mostly used to visualize the characteristics of clusters that emerge at the end of the analysis. They are used to plot one or more groups of values over multiple common variables. This is achieved by giving an axis for each variable, and these axes are arranged radially around a central point and spaced

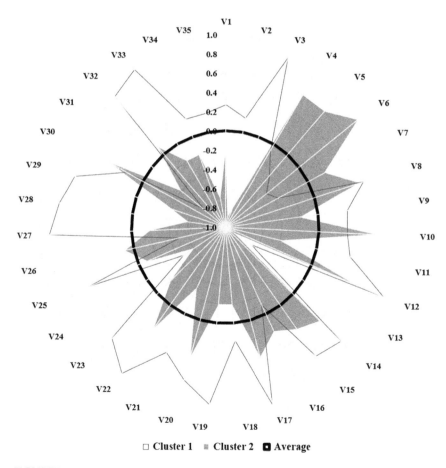

□ Cluster 1   ▪ Cluster 2   ◼ Average

FIGURE 7.3
Illustration of radial plot.

equally. Data from a single observation are then plotted along each axis and linked together to generate a polygon. As shown in Figure 7.3, multiple polygons can be placed within a single chart to contrast multiple observations. In this example, two geodemographic clusters are contrasted. The opacity of polygons has also been reduced to allow for easier visual interpretation.

Radial plots have numerous advantages that make them appealing to developers of geodemographic classifications. First, they are particularly useful for visualizing comparisons since multiple attributes can be easily contrasted along their own axis. Doing this ensures that overall differences are immediately apparent given the sizes and shapes of polygons. Radial plots also have the advantage of representing many variables side by side while giving each variable the same resolution. They further have the advantage of showing data outliers and commonality in a very striking manner.

Despite numerous advantages of radial plots, it is important to keep in mind that these visualizations also have some shortcomings. A key challenge with using radial plots is that they can generate some confusion once there are too many webs or polygons on the charts. This is because the incorporation of too many webs results in many axes, and ultimately crowded data. Sometimes, lowering the opacity of the polygons may help ameliorate this problem. Nevertheless, doing this does not eliminate the problem of distinguishing the shapes or colors of individual polygons. Therefore, it is good practice to keep radial plots simple and limit the number of variables used (Rahlf, 2017). Another shortcoming of radial plots is concerned with the possibility of erroneous interpretation. Sometimes when people look at radial plots, they could potentially assume that the area within the polygons is the most important thing to consider. However, the area within polygons is merely a reflection of the ordering of variables. Therefore, polygonal areas can change greatly depending on how the axes are positioned around the circle. For instance, in Figure 7.3, variables are arranged in ascending order of the variable code (v1 to v35). Should the ordering change to a descending arrangement of the variable codes, the shape of the polygons within the chart will also change.

7.3 Node-Link Trees

Within the context of machine learning literature, a tree is fundamentally a graph in which any two nodes are connected by exactly one path (Chen et al., 2008). Four major tree visualizations are popularized within machine learning literature. These include (a) indented lists, (b) layered diagrams, (c) tree maps, and (d) node-link trees, which are illustrated in Figure 7.4. Node-link trees are the most popular types of tree diagrams embraced within the field of geodemographics.

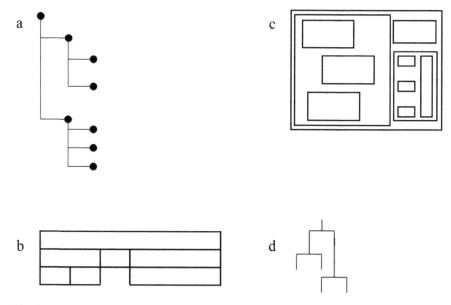

FIGURE 7.4
Prominent examples of tree visualizations.

Node-link trees (also called dendrograms) are used for visualizing hierarchical data in a tree-like structure. These tree plots are particularly useful for illustrating compound relationships. Therefore, they are commonly used to present outputs from hierarchical machine learning algorithms (Chen et al., 2008). The end result of a hierarchical clustering algorithm is a sequence of nested and indexed partitions. This makes node-link tree plots well suited for visualizing these partitions. These visualizations are used to figure out the appropriate way to allocate objects located in taxonomic space into clusters. Node-link tree plots display clusters, subclusters, relationships, and the order in which objects are merged. A primary concern of node-link trees is the spatial arrangement of nodes and links. Often (but not always) the goal is to effectively depict connectivity patterns, partitions, clusters, and outliers as shown in Figure 7.5, which depicts the hierarchical clustering of 32 African countries.

One reason as to why the natural visualization method for hierarchical clustering is the node-link tree plot is that it directly reflects the construction principle of the underlying algorithm. However, the resulting output must be interpreted with care. The key to reading a node-link tree plot is to focus on the height at which any two objects are joined together. For instance, in Figure 7.5, one can see that Liberia and Sierra Leone are most similar because the height of the link that joins them together is the smallest.

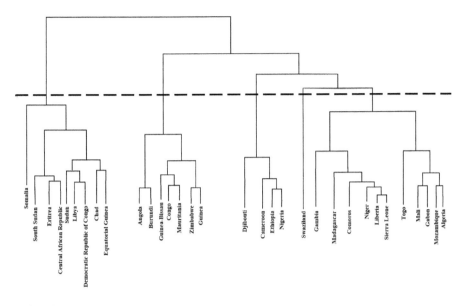

FIGURE 7.5
Illustration of dendrogram.

7.4 Bar and Column Plots

Bar and column plots display statistical data using rectangular bars where the length of the bar is proportional to the data value being represented. Both types of charts are used to compare multiple values. However, the difference between them lies in their orientation. The bar plot is oriented horizontally, while the column plot is oriented vertically. These plots are used to display categorical data. One axis of the plot shows the specific categories being compared, and the other axis represents a measured value. Both bar and column plots allow for pictorial renditions of statistical data so that independent variables can attain only certain discrete values (Rahlf, 2017).

In the field of geodemographics, bar and column charts are mostly used to distinguish positive values from negative values, as illustrated in Figure 7.6. As shown in the figure, categories displaying traits above a certain level (e.g. the average) point to the right, while those categories with traits beneath the mean point to the left.

Bar plots are mostly used to support geodemographic area-based profiling. Profiling refers to the systematic identification of characteristics that differentiate members of a target population (Harris et al., 2005), for example, your customers from people who are unlikely to ever buy your product or have a requirement of your service. Profiling is needed to target marketing

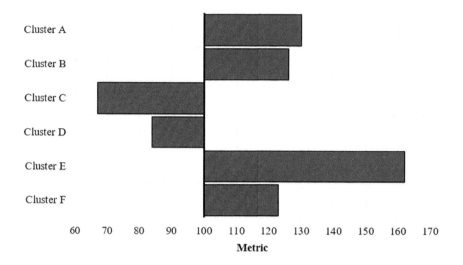

FIGURE 7.6
Illustration of bar plots.

and intervention campaigns, evaluate market potential, develop appropriately balanced sales and franchise territory areas, identify where to locate new outlets, identify hot spots to target sales effort, and to help tailor public communications.

7.5 Area Cartograms

One of the main objectives of spatial modeling and GIS is to represent reality as accurately as possible. However, area cartograms are used to accomplish the direct opposite of this objective. Rather than mimic the reality, area cartograms are used to distort reality while presenting information. Area cartograms are visualizations that describe attributes of geographic objects based on the resizing and exaggerating of the areas (Dorling, 2012). They are not "true" maps because they do not depict geographic space; rather, they alter the sizes of corresponding areas.

In GIS literature, area cartograms are broadly classified into three categories based on the way they show the attributes of geographic objects: noncontiguous, contiguous, and Dorling cartograms. For noncontiguous cartograms, objects are freed from their neighbors and therefore there is no need of the output to maintain the topological qualities of the inputs.

Unlike noncontiguous cartograms that sacrifice topology, contiguous cartograms maintain topology. Since topology has to be maintained, this creates a dual challenge for the cartographer. First, the cartographer needs to ensure that the geographic areas being distorted are transformed into sizes that correspond with their attribute values. Second, the cartographer must seek to maintain the shapes of these objects as much as possible. This dual challenge makes the creation of contiguous cartograms more difficult than their noncontiguous counterparts. The third type of cartogram, the Dorling cartogram, maintains neither shape topology nor object centroids, and it has proven to be an effective cartogram method.

To illustrate the benefits of visualizing social spatial structure with cartograms, Figure 7.7 is used to contrast two maps showing the same variable. The map on the left shows the geographic view that is the undistorted configuration of map areas. This map shows the actual size of the geographic units mapped. The second map (proportional view) is a contiguous area cartogram called hexograms. These types of hexograms can be produced through a procedure called hexagonal binning (Harris et al., 2018). In the proportional view shown in Figure 7.7, all geographic units have been transformed into the same size, and as can be seen, the original shape of the map has been maintained as much as possible.

As shown in Figure 7.7, in the geographic view, those geographic units that are quite small in size are obscured. Harris et al. (2018) therefore suggested that an obvious solution for this problem is to make obscured geographic units bigger. In the proportional view, hexograms have been used to enlarge the sizes of small geographic units, thereby allowing the readers to clearly see and visually interpret the true spatial configuration of the variable at a glance.

7.6 Online and Offline Interactive Visualizations

Interactive visualizations can also be useful for exploring and interpreting outputs of small area geodemographic data and information products. Online interactive visualizations such as mashups often require access to the Internet. A mashup is a technique by which a website or web application uses data, presentation, or functionality from two or more sources to create a new service (Gibin, 2009). Mashups are made possible via web services or public Application Programming Interface (APIs) that generally allow free access. The main characteristics of the mashup are combination, visualization, and aggregation. Most mashups are visual and interactive in nature as illustrated in Figure 7.8.

Small area geodemographic data and information products can also be made available in other formats like compact discs and data sticks to enable

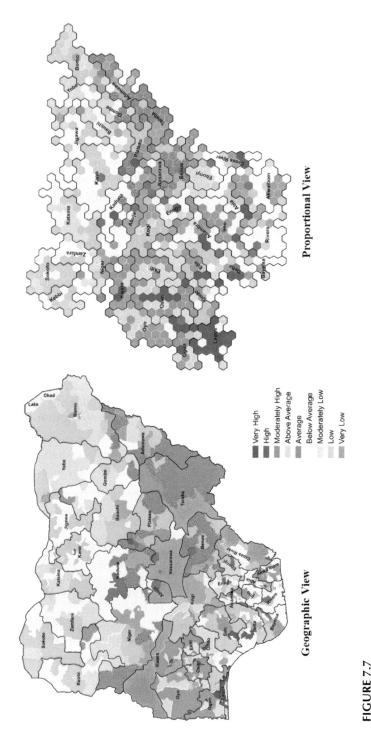

FIGURE 7.7
Illustration of undistorted map *vs* area cartograms.

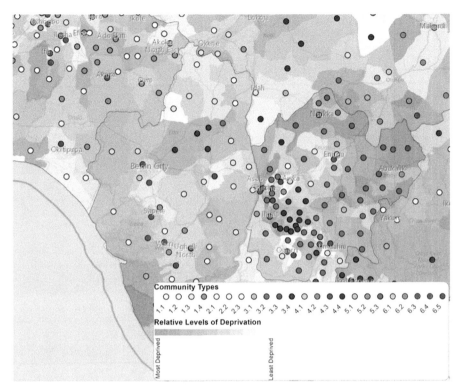

FIGURE 7.8
Google Openstreetmap mashup illustration of geodemographic information system.

those without regular access to the Internet to use the products. An example of an offline multimedia toolkit is shown in Figure 7.9.

7.7 Conclusion

The development and deployment of small area classifications entail the interrogation of large spatial data sets. Analytical procedures often generate volumes of multidimensional outputs that need to be effectively presented and interpreted by the human mind. Adequate graphic displays are therefore important for effective communication and understanding of outputs of the analysis. As demonstrated in this chapter, the visualization goals of small area geodemographic classification products are multifaceted. The choice of visualization methods therefore needs to be appraised by making nuanced judgments as demonstrated in this chapter. Some prominent examples of static

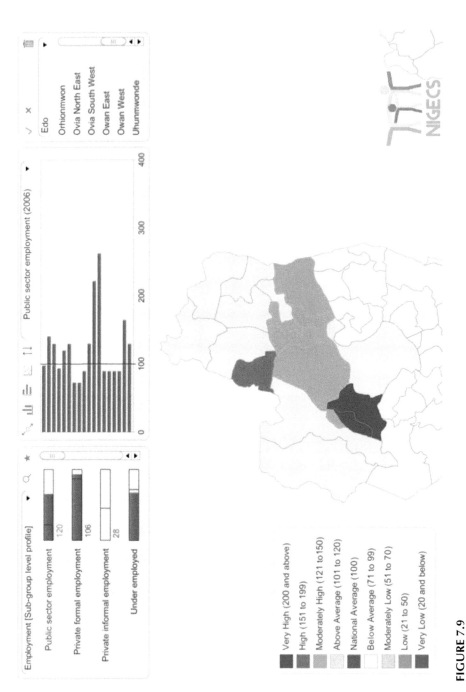

FIGURE 7.9
Example of an offline interactive multimedia toolkit.

and dynamic visualization techniques used in the field of geodemographics have been discussed here. Although the list is not exhaustive, it is hoped that the discussion will help to raise the profile of the importance of careful thinking about visualization methods when trying to present compelling geodemographic information beyond the corridors of technical audiences.

References

Chen, C., Härdle, W. and Unwin, A. (Eds). (2008). *Handbook of Data Visualization.* Berlin: Springer.

Dorling, D. (2012). *The Visualization of Spatial Social Structure.* Chichester: John Wiley & Sons Ltd.

Galbraith, R.F. (1994). Some Applications of Radial Plots. *Journal of the American Statistical Association*, 89(428), 1232–1242.

Gibin, M., Mateos, P., Petersen, J. and Atkinson, P. (2009). Google Maps Mashups for Local Public Health Service Planning. In: S. Geertman and J. Stillwell (Eds), *Planning Support Systems Best Practice and New Methods.* Berlin: Springer.

Harris, R., Charlton, M. and Brunsdon, C. (2018). Using Hexograms to Map Areal Data. In: *Proceedings of the 26th GIS Research UK Conference*, Leicester, Newcastle, 23 April 2019 - 26 April 2019. Available at: http://gisruk.org/ProceedingsGISRUK2018/GISRUK2018_Contribution_023.pdf.

Harris, R., Sleight, P. and Webber, R. (2005). *Geodemographics, GIS and Neighborhood Targeting.* London: Wiley.

Leventhal, B. (2016). *Geodemographics for Marketers: Using Location Analysis for Research and Marketing.* London: Kogan Page.

Rahlf, T. (2017). *Data Visualization with R: 100 Examples.* Berlin: Springer.

Rom, M.C. (2015). Numbers, Pictures, and Politics: Teaching Research Methods through Data Visualizations. *Journal of Political Science Education*, 11(1), 11–27.

Vickers, D.W. (2006). *Multi-Level Integrated Classifications Based on the 2001 Census.* Unpublished PhD thesis. School of Geography, University of Leeds.

Webber, R. and Burrows, R. (2018). *The Predictive Postcode: The Geodemographic Classification of British Society.* Los Angeles, CA: Sage.

Part 3

Illustrative Applications
and Conclusion

8

The Grouping of Nigerian Local Government Areas

8.1 Nigeria: Structure of Administrative, Census, and Electoral Geographies

Estimates derived from the 2019 UN World Population Prospects indicate that Nigeria's current population has climbed up to 206 million people in 2020 (UN, 2019). Nigeria has maintained its status as the most populous country in Africa since its independence from Britain in 1960. Between 2015 and 2020, its population growth averaged 2.6% (UN, 2019). During the same period, the global average population growth rate was 1.09%, making Nigeria one of the fastest growing countries in the world. Ojo and Ojewale (2019) have explained in their recent work that high fertility rates have contributed to Nigeria's increasing population particularly within urban centers.

One of the striking features of Nigeria's present demographic profile is that a youth bulge has emerged in the country. This has long been foreseen by economic demographers. As shown in Figure 8.1, estimates calculated by the UN confirm that the median age of the global population is rising, suggesting that the world population is aging (UN, 2019). However, there is a yawning gap between the median population in Africa and the rest of the world. The African region presently has the largest share of young people with a median age of 19.68 years as of 2020. Nigeria's median age has fallen from 19.11 years in 1950 to 18.06 years in 2020.

Comprising an area totaling 356,669 square miles (Gordon, 2003), Nigeria has a rich base of human capital and natural resources. Petroleum is the main source of foreign exchange. However, due to repeated violent conflicts in the oil-rich Niger-Delta region of Nigeria, oil production in Nigeria has diminished significantly, causing the country to drop out of the top ten (now ranked 13th globally) oil-producing nations in the world. Nigeria maintains a high-profile economic and political status on the African continent. Indeed, Gordon (2003) has explained that several African countries have their economic stability hooked to the political and economic stability of Nigeria.

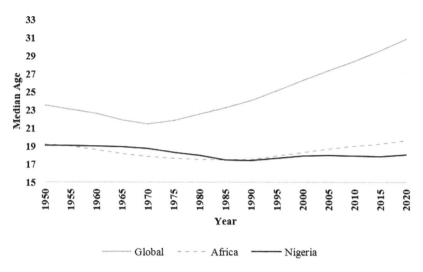

FIGURE 8.1

Median age in Nigeria versus Africa and the world (Author's elaboration based on data from the United Nations, Department of Economic and Social Affairs, Population Division (2019). World Population Prospects 2019, Online Edition).

Nigeria is made up of more than 250 ethnic groups, each with its local language (dialect), and multiple religions. Among these groups there are three dominant ethnic nationalities that have the largest populations. The Yorubas are located mainly in the southwestern corner of the country. The Igbos can be found mainly in the southeastern part of the country, while the Hausa–Fulani ethnic groups are concentrated in the northern part of the country.

Before pressing ahead with the rest of this chapter, it is important to give a brief overview of the structure of Nigeria's key administrative, census, and electoral geographies. The structure of these geographies is not very complicated as shown in Figure 8.2. Due to frequent modifications, the number of divisions at different geographic levels has changed from time to time. All the numbers shown represent geographic divisions as of 2019. However, the number of enumeration areas (EAs) shown corresponds to 2006, which was the last time a national census was conducted in Nigeria.

This chapter focuses on the administrative geography of Nigeria which is also very much linked to the spatial configuration of the country's ethnic distribution. Nigeria's top level administrative geography comprises six geopolitical zones, as shown in Figure 8.3. The southwest geopolitical zone is dominated by the Yoruba ethnic group. The Igbos mainly reside within the southeast geopolitical zone, and the Hausa and Fulani ethnic groups can be found mainly within the northeast and northwest geopolitical zones. The

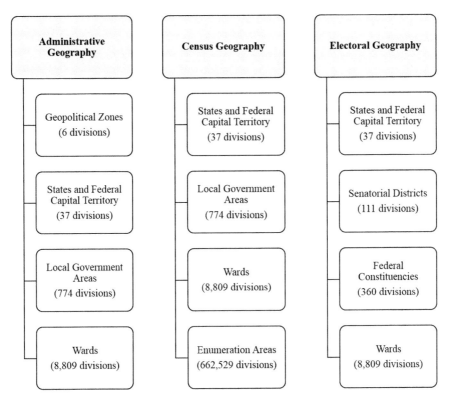

FIGURE 8.2
Hierarchical structures of Nigeria's administrative, census, and electoral geographies.

north central and south-south geopolitical zones are interesting melting pots of different ethnic groups, with no particular ethnic group exhibiting dominance in terms of total population representation.

The six geopolitical zones are subdivided into 37 subdivisions [36 states and 1 Federal Capital Territory (FCT) named Abuja]. Each state is administered by a state government, with a governor as the head. States are further divided into 774 local government areas (LGAs). The LGAs are important for ensuring that national policies reach the grass roots. The constitutional responsibilities of LGA detailed in the Constitution of the Federal Republic of Nigeria 1999 include:

- Making economic recommendations to the state
- Collection of taxes and fees
- Establishment and maintenance of cemeteries, burial grounds, and homes for the destitute or infirm

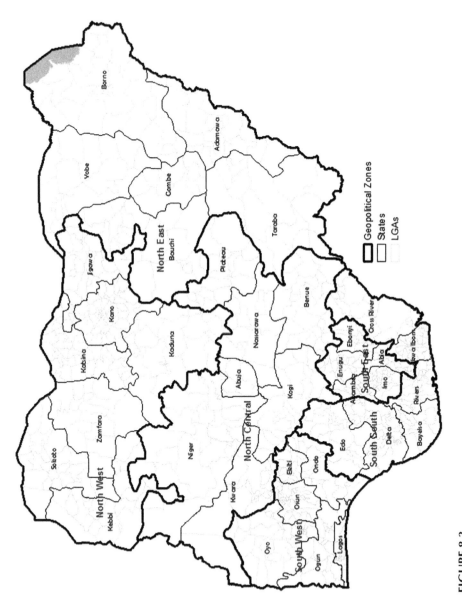

FIGURE 8.3
Administrative geography of Nigeria.

- Licensing of bicycles, trucks (other than mechanically propelled trucks), canoes, wheel barrows, and carts
- Establishment, maintenance, and regulation of markets, motor parks, and public conveniences
- Construction and maintenance of roads, streets, drains, and other public highways, parks, and open spaces
- Naming of roads and streets and numbering of houses
- Provision and maintenance of public transportation and refuse disposal
- Registration of births, deaths, and marriages
- Assessment of privately owned houses or tenements for the purpose of levying such rates as may be prescribed by the House of Assembly of a state
- Control and regulation of outdoor advertising, movement and keeping of pets of all descriptions, shops and kiosks, restaurants, and other places for sale of food to the public, and laundries.

The LGA administrative level is the geographic scale at which national and state governments expect the impacts of some of their policies to reach people at the grass roots (Ojo and Ezepue, 2012). Unfortunately, little countrywide spatial analysis has been conducted at this scale. Using data from Nigeria's 2006 Census and the Core Welfare Indicators Questionnaire (CWIQ) survey supplied by the National Bureau of Statistics (NBS), the remainder of this chapter summarizes how spatial analysis and unsupervised machine learning have been used to group all 774 LGAs. The chapter provides a review of the reasoning behind the variables selected and clarifies the analytical procedures used to derive the area classification system.

8.2 Data and Input Variables

One of the reasons for the paucity of local level spatial analysis in Nigeria has been the difficulty associated with accessing digital statistical data sets for Nigerian LGAs, Wards and EAs. The National Population Commission (NPC) has a statutory responsibility for collecting and publishing census data. However, over the years, the NPC has struggled with capacity to publish these data sets in digital and accessible formats (Turnwait and Odeyemi, 2017). Furthermore, the NPC has not been releasing small area population projections. The weakness of this statutory agency coupled with several unnecessary bureaucratic bottlenecks has greatly undermined the conduct of small area spatial analysis in Nigeria.

Following a merger of the Federal Office of Statistics (FOS) and the National Data Bank (NDB), a new agency, the NBS, was established in 2007. The NBS has been more responsive to the increasing statistical needs of researchers and policymakers who wish to conduct socioeconomic research in Nigeria. The NBS coordinates nationwide statistical surveys and generates official statistics across all federal ministries, departments and agencies (MDAs), state statistical agencies (SSAs) and local government councils (LGCs). The NBS supplied the data sets used in the study reported in this chapter. The data sets were derived from two sources:

- The 2006 Population and Housing Census
- The 2006 CWIQ survey

The initial data set from both data sources comprised 644 variables for all 774 LGAs, resulting in 498,454 data points. The full data set was manually explored, and as a starting point, ten data themes were identified. The ten data themes observed within the data set include:

- Agriculture
- Demographics
- Education
- Employment
- Health
- Household composition
- Household infrastructure
- Housing
- Socioeconomics
- Women and children.

Although the 644 variables were assigned to one of each theme, it was impossible to use all the variables for multiple reasons as detailed in Chapter 5 of this book. An initial representative number of variables were considered. Variables with multiple missing values were avoided when pooling together an initial list from the initial 644 variables. Although artificial data generation techniques, such as microsimulation, can be used to fill the missing data (Ballas et al., 2005), it was decided that synthetic data sets should be avoided given that this was the first geodemographic area classification developed for the country. Avoiding synthetic data should prevent or minimize possible distortions to the output. In addition to excluding variables with several missing data, the policy relevance of variables was also considered when making the initial selection of variables. An initial selection of 125 variables was made from the original list of 644 variables for further statistical evaluation.

8.2.1 Statistical Approaches for Variable Reduction

The 125 initial variables were reduced in two stages. First, a theme-by-theme data preprocessing was done. For instance, all the variables within the education theme were tested against themselves using various statistical techniques and principles as detailed in Section 5.2. This intra-theme data preprocessing resulted in the first phase of data reduction. Variables from all ten themes that survived the first phase of data preprocessing were then tested against themselves (inter-theme data preprocessing). This dual-stage data preprocessing approach helped to ensure that the data reduction procedure was rigorous.

The proportion of the population for which each variable accounts was considered. During data preprocessing, variables with small sample sizes were avoided because such variables have a tendency to be volatile and they change rapidly over time. Such variables would not sustain the longevity of area classification.

Table 8.1 shows the top ten variables exhibiting the smallest sample proportions across LGAs. The percentage share of households that built their residences with cardboard is ranked number one. Averagely, the variable accounts for only 0.48% of households within each LGA. Another problem that can result from variables with small sample proportions is that they provide little distinctive information for naming and profiling the groups that emerge at the end of the clustering process. A solution considered for some of the variables with small sample sizes was to merge them (where they share a similar base). For instance, the variable *% children not in school due to early marriage* was merged with the variable *% children not in school due to teenage pregnancy* because they share the same base (100%) and fall within the same theme (women and children).

In addition to the evaluation of the sample proportions represented by variables, skew was also analyzed. Positively skew occurs because of

TABLE 8.1
Variables with the smallest sample sizes

Variable	Rank
% of households built with cardboard	1
% of children not in school due to teenage pregnancy	2
% of households built with stone	3
% of children not in school due to early marriage	4
% of households built with iron sheets	5
% of duplexes	6
% of households built with wood/bamboo	7
% of subsidized occupancy status	8
% of children under 5 not breastfed	9
% of population employed in the fishing industry	10

Source: Author's calculations.

TABLE 8.2

Variables with the largest positive skew

Variable	Skew
% of households which always find it difficult to pay house rent	11.25
% of households built with cardboard	11.15
% of children not in school due to teenage pregnancy	8.87
Populati on density	8.09
% of households built with stone	7.41

Source: Author's calculation previously published in Ojo, A., Vickers, D. and Ballas, D. (2012). The Segmentation of Local Government Areas: Creating a New Geography of Nigeria. *Applied Spatial Analysis and Policy*, 5(1), 25–49.

an accumulation of large values at the lower end of a variable distribution or where there are outliers or extreme values within the distribution. Variables that exhibited positive skew were mostly deselected during data preprocessing. Table 8.2 shows some of the variables with the largest positive skew. The problem with most of these variables is that they identify small proportions of the total population. Consequently, they concentrate at the lower end of the 0 to 100% scale. Vickers and Rees (2006) found that variables that would work well within the classification are those that spread in their variation across geographic areas.

Consideration was also given to the associations between variables within the data set. The inclusion of highly correlated variables in an unsupervised learning algorithm will result in the duplication of the same information or population attribute. This can give undue advantage to that population attribute and mask other important attributes of the population. The duplication of information in this manner is called redundancy.

Pearson's product moment correlation co-efficient was used as the statistic for evaluating the association between variables. High positive and negative correlations are not desirable for unsupervised machine learning algorithms because they contribute to redundancy. Some high correlations were observed within the Nigerian data set. Table 8.3 shows the correlation matrix of variables within the education theme.

Values in orange color in Table 8.3 indicate high positive correlations, while those in blue indicate high negative correlations. People who are uneducated have a high negative correlation with people who completed secondary education. The reason for this is that these variables share the same denominator and an individual can only belong to one of the two categories. Cross-correlation was also evaluated for variables in different themes. Table 8.4 shows the results of correlation analysis for some variables within different themes that do not share the same denominator.

The types of associations among variables shown in Table 8.4 are due to the ability of one variable to explain the variation existing within the other variable. This implies that the attributes of the population being reported by one

TABLE 8.3
Correlation matrix of variables within the education theme

E1	E2	E3	E4	E5	E6	E7	E8	E9	E10	
	−0.77	−0.82	−0.65	0.95	−0.77	−0.01	−0.19	−0.25	−0.31	E1
		0.47	0.24	−0.72	0.57	−0.15	0.17	0.06	0.15	E2
			0.71	−0.83	0.73	0.05	0.17	0.32	0.32	E3
				−0.67	0.62	0.19	0.1	0.44	0.34	E4
					−0.78	0.01	−0.21	−0.27	−0.3	E5
						0.06	0.18	0.3	0.32	E6
							−0.51	0.63	0.27	E7
								−0.27	0.27	E8
									0.22	E9
										E10

Source: Author's calculation.

E1: Uneducated population.
E2: Population that completed primary school.
E3: Population that completed secondary school.
E4: Population with postsecondary education.
E5: Head of household uneducated.
E6: Adult literacy rate.
E7: Under 15 minutes to nearest primary school.
E8: 15 to 29 minutes to nearest primary school.
E9: Under 15 minutes to nearest secondary school.
E10: 15 to 29 minutes to nearest secondary school.

variable can also be caused by the presence of the other variable. For instance, the variable for people who own two to ten cattle has a high positive correlation (0.82) with the variable for people who are not educated. This is because across Nigeria, literacy rates are generally lower within those communities that have large concentrations of people whose primary occupation is cattle farming (Federal Government of Nigeria, 1987; Iro, 2006). The positive nature of the correlation indicates that there is a higher tendency for people who are not educated and own cattle to live within the same community.

It is also important that variables spread well in order for them to perform well in the unsupervised clustering algorithm. A useful statistic for measuring the geographic variation of variables is the standard deviation. Table 8.5 shows those variables displaying high levels of geographic spread. The percentage of people who own motorcycles combines a mean of 31.98 with a standard deviation of 28.56. From this, we can deduce that two-thirds of the values of the variable lie between 3.42 and 60.55.

Variables with large standard deviations will prove more useful than those with lower values because they present better distinctions between areas (Harris et al., 2005). However, some variables with lower standard deviations may be merged with other variables and renamed if they share the same

TABLE 8.4
Relationship and Redundancy among selected variables

Variable	Theme	Variable	Theme	Correlation	Redundancy (%)
Owns 2–10 cattle	A	Uneducated population	E	0.82	67
Never married	D	Uneducated population	E	−0.78	61
Postsecondary education	E	Ownership of mobile phones	S	0.78	60
Renting a house	H	Difficulty in paying house rent sometimes	S	0.77	59
Owns 2–10 cattle	A	Uneducated household head	E	0.75	57
Uneducated household head	E	Built with cement/sandcrete	HO	−0.75	56
Postsecondary education	E	Nonwood fuel for cooking	HI	0.74	55
Private formal employment	EM	Ownership of mobile phones	HI	0.74	54
Postsecondary education	E	Renting a house	HO	0.72	52
Age 0–14 years	D	Completed secondary education	E	−0.71	51
Never married	D	Uneducated household head	E	−0.71	51

Source: Author's calculation previously published in Ojo, A., Vickers, D. and Ballas, D. (2012). The Segmentation of Local Government Areas: Creating a New Geography of Nigeria. *Applied Spatial Analysis and Policy*, 5(1), 25–49.

A: Agriculture
D: Demographics
E: Education
EM: Employment
H: Health
HC: Household composition
HI: Household infrastructure
HO: Housing
S: Socioeconomics
WC: Women and children

TABLE 8.5
Variables with high standard deviations

Variable	Mean	Standard Deviation
% of households built with cement/sandcrete	40.14	33.22
% of households built with mud/mud bricks	54.45	31.80
% of households using agricultural inputs	40.70	30.74
% of households with access to safe toilet sanitation	50.68	28.96
% of households owning motorcycles	31.98	28.56
% of households with access to safe sources of drinking water	45.63	28.56
% of single room housing unit	66.52	28.13
% of whole buildings	26.87	27.98
% of the total population uneducated	39.60	26.55
% of households owning less than 1 hectare of land	26.42	26.09

Source: Author's calculation previously published in Ojo, A., Vickers, D. and Ballas, D. (2012). The Segmentation of Local Government Areas: Creating a New Geography of Nigeria. *Applied Spatial Analysis and Policy*, 5(1), 25–49.

base. A new variable that was created to improve its standard deviation is the variable *cattle ownership*. The variable *percentage of people who own over 50 cattle* displayed little variation across LGAs (standard deviation = 1.42). Consequently, this variable was merged with three other variables: *the percentage of people who own 2 to 10 cattle*, the *percentage of people who own 11 to 20 cattle*, and *the percentage of people who own 21 to 50 cattle*. These variables were merged to create a new variable called *cattle ownership*. Consequently, the standard deviation of the new variable increased to 22.19.

8.2.2 The Final List of Variables

Making decisions about the final choice of variables is an important activity. At the start of data preprocessing, there were 125 variables spread across 10 data themes. The socioeconomics data theme had the largest share of variables (29%), while the health theme accounted for just 2% of all variables at the start of data preprocessing. At the end of the intra-theme data reduction process, the socioeconomics theme accounted for 20% of all variables. Following the inter-theme data reduction process, the household infrastructure theme moved into the lead, accounting for 18% of all variables. Agriculture, health and household composition themes remained the least represented. In a nutshell, data preprocessing commenced with 125 variables that were reduced to 54. These were further reduced to a final list of 45 variables, as shown in Table 8.6.

8.2.2.1 Agriculture Theme and Variables

The agricultural sector plays a major role in Nigeria's socioeconomic development and thus cannot be underestimated. Not only is agriculture important for food supply, the sector also has the potential to generate

TABLE 8.6
The final list of 45 variables

Variable	Data Theme
Use of agricultural inputs	Agriculture
Owns cattle	Agriculture
Age 0–14 years	Demographics
Age 15–59 years	Demographics
Age 60 years and over	Demographics
Never married	Demographics
At least one pensioner	Demographics
Population density	Demographics
Separated couples	Demographics
Completed secondary education	Education
Head of household uneducated	Education
Adult literacy rate	Education
Access to primary school	Education
Economically active population	Employment
Self-employed	Employment
Employment in the transport sector	Employment
Employment in agriculture	Employment
Taking anti-malaria measures	Health
Access to health	Health
Household size 1–2 persons	Household composition
Mean household size	Household composition
Ownership of mobile phone	Household infrastructure
Ownership of personal computer	Household infrastructure
Access to safe source of water	Household infrastructure
No toilet facility	Household infrastructure
Safe toilet sanitation	Household infrastructure
Nonwood fuel for cooking	Household infrastructure
Lighting energy-mains electricity	Household infrastructure
Access to water supply	Household infrastructure
Own a house	Housing
Renting a house	Housing
Free accommodation	Housing
Built with burnt bricks	Housing
Single room	Housing
Duplex	Housing
Built with cement/mud brick	Housing
Vehicle ownership	Socioeconomics
Motorcycle ownership	Socioeconomics
Size of land more than 6 hectares	Socioeconomics
Improved security	Socioeconomics
Access to transportation	Socioeconomics
Difficulty in meeting basic needs	Socioeconomics
Children living with single parents	Women and children
Early marriage and teenage pregnancy	Women and children
Vaccinated	Women and children

Source: Author's compilation previously published in Ojo, A., Vickers, D. and Ballas, D. (2012).
The Segmentation of Local Government Areas: Creating a New Geography of Nigeria.
Applied Spatial Analysis and Policy, 5(1), 25–49.

large-scale employment opportunities especially for Nigeria's rural population (Adesugba and Mavrotas, 2016). The agricultural sector generates
raw materials for the manufacturing sector. In addition, agricultural cash
crops are key sources of foreign exchange. Furthermore, the eradication of
extreme hunger and poverty in Nigeria requires a vibrant agricultural sector
(Nwankpa, 2017). Variables within this data theme can also present interesting
geographic information on Nigeria's urban–rural divide. In addition, agricultural variables can indicate parts of the country that are food deserts.

8.2.2.2 Demographics Theme and Variables

Demographic studies explore issues relating to the size, structure,
dispersement, and development of human populations (Bloom et al., 1998).
Age captures attributes of individuals during different periods in their life
cycle. Age can also help to shed light on health-related issues, particularly
sexual health (House et al., 1990). Marital information can be a very useful
proxy in fertility studies. The geographic variation of marital status has an
implication on growth. Marriage can have a financial implication for individuals (Morgan, 1991). In addition, divorce rates have economic and social
impacts (Morgan, 1991). Retirement after active service can result in a change
of lifestyle. There is a general trend in Nigeria for people to retire back to
their hometowns after active service. The pensioners group was included
because it is important for planning and development, as they constitute a
great percentage of the economically inactive population. Population density
helps to distinguish urban from rural areas. High population density can
also trigger social pressures on housing and other public amenities.

8.2.2.3 Education Theme and Variables

Education plays a major role in the measurement of poverty prevalence.
Evidence has shown that the incidence of poverty decreases with increasing
level of education in sub-Saharan Africa (Palmer et al., 2007). For instance,
those who have completed secondary education and above add to the economically active population. The role of the household head in Nigeria is very
important in defining access to education. Less educated household heads
are less likely to generate high incomes (NBS, 2006) and may be exposed to
poverty. Adult literacy gives a sense of those areas where older population
groups crave for knowledge. Information about geographic access to educational institutions can also provide a better understanding of how to plan
properly for investment in the educational sector.

8.2.2.4 Employment Theme and Variables

Employment is usually the key source of income for most people; therefore,
unemployment aggravates poverty. Two principal sectors of employment

have dominated the Nigerian labor market. The public sector employs people who work mainly in government departments, while the private sector is for nongovernment workers. However, since Nigeria has a massive youth population, there is a yawning gap between the available number of jobs and the number of unemployed. In order to meet the demands of the growing workforces, Nigeria needs to create millions of jobs annually over the next decade. The public and private sectors alone cannot possibly fill this gap. Therefore, there is increasing expansion across Nigeria's voluntary sector. Different variables are included to capture the spatial divisions of economic activity and productivity. A person is economically active when he or she engages in economically productive work. This variable is a variant, which combines a proportion of the employed and unemployed population. The variable showed reliable variation across areas. Self-employment was also used because of its greater variation across spaces. A composite variable was formed by merging the population employed in the agricultural sector with those in the fishing sector, as both constitute agricultural work. Transport sector employment is indicative of areas where there is a reasonable level of utilization of public transportation.

8.2.2.5 Health Theme and Variables

Individual and collective health can be heavily influenced by the dynamics of geography. There is an inherent link between people's birth place, where they reside, study, work, socialize, and ultimately die. Places affect health experiences in diverse ways. Spatial location also plays a fundamental role in shaping social and environmental risks that inform health outcomes. Two health variables were included in the final list of variables: the share of *persons taking anti-malaria measures* and a measure of *access to healthcare*. The fight against malaria is a key priority of the Ministry of Health in Nigeria. In 2018, Nigeria and five other countries (Democratic Republic of the Congo, Uganda, Côte d'Ivoire, Mozambique, and Niger) accounted for more than half of all malaria cases globally. Approximately 25% of global malaria cases in 2018 were reported in Nigeria (WHO, 2019). This disease exerts a huge financial burden on Nigeria's healthcare purse. Preventive measures against malaria are therefore not only important for health and well-being, but also significant in terms of their economic benefits. The variable measuring access to health captures travel time. It is defined as the share of the population who travel up to 30 minutes to their nearest healthcare facility. This variable was created by merging two other variables (*share of population who travel less than 15 minutes to the nearest health facility* and *those who travel between 15 and 29 minutes*).

8.2.2.6 Household Composition Theme and Variables

Although the economic benefits of large household sizes remain debatable in Nigeria, household composition and size have economic implications.

Within social settings, children tend to make significant economic contribution to their parents in the long run (Hagen-Zanker and Holmes, 2012). In such circumstances, those households with many children should normally be better-off than those with fewer children. However, the first few years of a child's life affect his or her future outcomes as well as his or her future position in the income distribution. In Nigeria, adverse experiences in early childhood (which is often associated with large poor households) are associated with lower rates of employment in later life (Hagen-Zanker and Holmes, 2012). Two variables (*one to two persons households* and *mean household size*) are used to give a sense of household composition. Smaller households have lower median incomes. Nevertheless, larger households with one earner will also have lower income bases than those smaller households with multiple earners. Mean household size acts as a measure of density and it also gives a sense of the likely economic priorities of areas. For instance, the large presence of children in a household can have major implications for the demand and allocation of resources for education and healthcare.

8.2.2.7 Household Infrastructure Theme and Variables

Household infrastructure can serve as a proxy for measuring the quality of life. Basic infrastructure relating to food, water, and sanitation is primary, while others like access to telecommunications are secondary. The spread of mobile phones in Nigeria has been unprecedented since 1999 (Jagun et al., 2008). Effective telecommunication is also very important for small- and medium-sized enterprises. It saves travel time, and also has an indirect positive consequence on the environment as greenhouse gas emissions are reduced. The variable measuring ownership of mobile phones has a better variation across geographic areas in Nigeria and it combines this with a reasonably large sample size. Access to water is defined for people who spend less than 30 minutes travel to their nearest water source. Access to water is important as it relates to the spatial patterns of waterborne diseases. Other important environmental indicators (people with access to toilet facilities and those with access to safe sanitation facilities) were included as they provide useful information on public health. They both show good variation across Nigerian LGAs and have representative sample sizes. The energy sector in Nigeria has witnessed numerous challenges over the years.

8.2.2.8 Housing Theme and Variables

For almost half a century, Nigeria like many other African countries has been rapidly urbanizing (Ojo and Ojewale, 2019). Rapid urbanization generates pressure for the housing sector. The provision and affordability of housing across Nigeria remain particularly strained, especially within urban centers. The variable *percentage of people who own houses* is used here as a proxy for housing affordability. This variable showed very good variation across LGAs and it also has reliable sample sizes. The variable *percentage of people renting houses* also gives a sense of housing demand. These two variables alongside the variable *percentage of people in free accommodation* were included in the study. Other housing-related variables included in the analysis are *the type of materials used for housing construction*. These variables are useful proxies of deprivation and serve as important discriminators of regional identity.

8.2.2.9 Socioeconomic Theme and Variables

One of the most difficult challenges faced by Nigeria's development sector is poverty reduction (Ogunbodede, 2006). Poverty breeds other social problems such as corruption and crime. A composite variable (*difficulty in meeting basic needs*) was included in the analysis. This variable subsumes the proportion of people who reported that they always have difficulty in satisfying their food needs, paying school fees, paying house rent, paying utility bills, and paying for healthcare. The ownership of a private means of transportation is also a useful proxy for socioeconomic status. Additionally, the variable *percentage of the population who reported that they owned over six hectares of land* was included. This variable showed a good variation across geographic spaces and it is also considered as a traditional measure of socioeconomic status in Nigeria. Crime and public safety have direct links with socioeconomics (Ojo and Ojewale, 2019). Self-perceived levels of crime were used as a proxy for the burden of crime. Two variables (*the percentage of people who reported that crime level was better* and *those who reported it was much better*) were merged together to create a new variable called *improved security*. Another composite socioeconomic variable, *access to transportation*, was constructed by merging the percentage of people who spend less than 15 minutes travel to their nearest transport source with the percentage of people who spend 15 to 29 minutes travel to the nearest transport source.

8.2.2.10 Women and Children Theme and Variables

The rights and role of women in today's world have consistently been an issue of significance. Like most women all over the world, Nigerian women face a variety of legal, economic, and social constraints. For instance, the determination to reduce maternal and child deaths continues to attract attention from local and global stakeholders. A significant proportion of these deaths can be prevented by increasing investment in health services, female education, and greater political commitment (Okonofua, 2010). Some of these decisions depend on a better understanding of geographic variations in information about women and children. Three variables were included within the women and children theme: *the percentage of children living with a single parent, a composite of the level of occurrence of early marriage or teenage pregnancy among young girls*, and *the percentage of children on a vaccination program*.

8.3 The Clustering Process

Following the selection of the input variables, the next step was to prepare the data for classification. All the variables for the 774 LGAs were assembled into a single database and manually checked for any errors. Once this was done, the clustering process could commence.

It is inappropriate to run a clustering algorithm on a data set that consists of variables of different scales. Since the variables were characterized by different measurement units (e.g. ratios and percentages), it was necessary to rescale them prior to standardization. However, prior to standardization, the variables were transformed using a logarithmic transformation method. This was necessary because some of the variables combined small dispersions with large means. Log transformation can cope well with the heterogeneity of the variance existing within some of the variables.

The z-score standardization technique was used to readjust variables measured in different scales because of its ability to maintain a mean of zero for the standardized values and a standard deviation of one. With a mean of zero, distortions stemming from the central value of each variable were avoided.

8.3.1 Clustering Algorithm and the Choice of Number of Clusters

The taxonomic analysis of Nigerian LGAs is about identifying homogeneous areas based on multivariate data. This sort of analysis has never been conducted in Nigeria at the local scale. Consequently, there is no basis for comparing the effectiveness of various clustering techniques. A variety of unsupervised machine learning methods have been explained in detail in Chapter 6. *K*-means clustering method was found to be less computation-intensive and it also generates several associated results, which prove useful for further evaluation of the clusters produced.

Careful consideration was given to the potential benefits of using the *k*-means algorithm to create a hierarchical classification. A decision was made to create a three-tier classification of LGAs. The top tier of the classification would comprise fewer groups and would be useful for general data visualization. However, this top tier of the classification may not be ideal for in-depth analysis of population disparities. A second- or third-tier hierarchy that would generate more groups would prove better for understanding and mapping local disparities.

Prior to creating the hierarchical structure, different techniques were used to ascertain the number of natural clusters generated by the *k*-means algorithm. First, the elbow method proposed by Everitt et al. (2001) was adopted. The algorithm was run to create 15 cluster solutions. The average distance of each case from its cluster center was calculated and plotted against the number of clusters as shown in Figure 8.4.

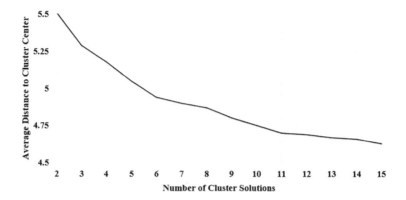

FIGURE 8.4

Distance from cluster centre versus number of clusters (Author's elaboration previously published in Ojo, A., Vickers, D. and Ballas, D. (2012). The Segmentation of Local Government Areas: Creating a New Geography of Nigeria. *Applied Spatial Analysis and Policy*, 5(1), 25–49).

A sharp change in the average distance to cluster center would suggest the optimal solution for number of clusters at the top hierarchy of the classification. Figure 8.4 suggests that the elbow is located around the six-cluster solution. Nevertheless, additional tests were used to confirm if the six-cluster solution was the optimal solution. Maps of the different cluster solutions were visually inspected, and it was found that the five-, six-, and seven-cluster solutions provided better distinctions between areas. These three solutions were subjected to additional statistical analysis.

The first statistical test evaluated the distance of each LGA from its cluster center, and was conducted for each of the three solutions. The lower the distance of a case from its cluster center, the better it is. The results of this test showed that the three different cluster solutions were positively skewed. The positive skew indicated that the majority of LGAs are under the lower distance categories. The larger the value of the positive skew, the better the solution in terms of how close cases are to the center of their cluster. The five- and seven-cluster solutions both yielded a skew of 0.31, while the six-cluster solution yielded a skew of 0.38. This test revealed that cases within the six-cluster solution are more compact than the other two solutions.

The second statistical test evaluated the membership sizes of each of the cluster solutions. In order to be useful for data profiling, an effective small area geodemographic classification should have a relatively balanced membership of cases. For the Nigerian LGA classification, cluster membership was evaluated by examining the range of the distribution for each of the three cluster solutions as shown in Table 8.7. The five- and six-cluster solutions outperformed the seven-cluster solution.

TABLE 8.7
Membership sizes of selected cluster solutions

		Cluster 1	Cluster 2	Cluster 3	Cluster 4	Cluster 5	Cluster 6	Cluster 7	Range
Five-Cluster	No. of LGAs	208	110	128	120	208			98
Solution	% of LGAs	26.9	14.2	16.5	15.5	26.9			12.7
Six-Cluster	No. of LGAs	181	166	114	126	82	105		99
Solution	% of LGAs	23.4	21.4	14.7	16.3	10.6	13.6		12.8
Seven-Cluster	No. of LGAs	107	102	162	160	52	89	102	110
Solution	% of LGAs	13.8	13.2	20.9	20.7	6.7	11.5	13.2	14.2

Source: Author's calculation previously published in Ojo, A., Vickers, D. and Ballas, D. (2012). The Segmentation of Local Government Areas: Creating a New Geography of Nigeria. *Applied Spatial Analysis and Policy*, 5(1), 25–49.

Sensitivity tests were also conducted on each of the three solutions to evaluate their response to uncertainty. The RMS error adapted by Gregory (2000) was described in Chapter 6 of this book. This statistic was used to evaluate the predictive power of the five-, six-, and seven-cluster solutions. To achieve this, actual LGA rates for seven indicators (shown in Table 8.8) were compared with rates predicted by the cluster grouping. For each of the three cluster solutions, the predicted values for the seven variables were subtracted from the actual values. The RMS was used to estimate the errors introduced. Results are shown in Table 8.8.

Results shown in Table 8.8 indicate that the five-cluster solution works with variables that are particularly correlated with the total population, which means that the six- and seven-cluster solutions are generally better in predicting those variables that are not necessarily correlated with the geographic distribution of the total population. It is challenging to draw precise conclusions between the performance of these two solutions. However, a closer observation of Table 8.8 shows that where the seven-cluster solution outperforms the six-cluster solution, the RMS error is marginal. Based on these findings, one can conclude that both the six- and seven-cluster solutions are less likely to generate large prediction errors. A final test of discriminatory power was deployed using the Gini coefficient (Brown, 1994). This test did not yield any significant differences in the performance of each of the three different cluster solutions.

After carefully considering the results generated from the statistical tests, the six-cluster solution was chosen as it outperformed the two other solutions. The selection of a six-cluster solution at the top hierarchy implied that the second and third hierarchies could be created. To create the second hierarchy, the top-down method used by Vickers (2006) was adopted. This method ensured that cluster groups in lower hierarchies maintain as much

TABLE 8.8
Results from RMS test

Indicator	Correlation with Total Population	RMS for Five-Cluster Solution	RMS for Six-Cluster Solution	RMS for Seven-Cluster Solution
Population in monogamous marriage	0.84	0.51	0.52	0.51
Population in public sector employment	0.67	2.56	2.45	2.43
Economically inactive population aged 15 to 24 years	0.88	3.24	3.25	3.38
Uneducated population	0.37	1.42	1.37	1.37
Household heads self-employed in agriculture	0.19	5.31	5.23	5.55
Population of children living in nonnuclear households	0.63	1.98	1.89	1.89
Children aged 0 to 4 years vaccinated for measles	0.88	2.98	2.74	2.79

Source: Author's calculation previously published in Ojo, A., Vickers, D. and Ballas, D. (2012). The Segmentation of Local Government Areas: Creating a New Geography of Nigeria. *Applied Spatial Analysis and Policy*, 5(1), 25–49.

traits as that of their parent clusters. Each of the six cluster groups identified at the first level were clustered separately using *k*-means algorithm. At the second level, between two and five cluster solutions were created and evaluated in a similar manner as the first level. The process was rigorous, and it helped to ensure that the results are robust. At the end of the analysis, 23 clusters were generated at the second hierarchy, and the third hierarchy yielded 57 clusters.

8.4 Labeling, Cluster Profiles, and Visualizations

The outputs of the Nigerian analysis have been interpreted in a number of ways. Maps, charts, and textual description have been used to communicate the complex results. First, each of the clusters has been given a label. Labeling a group of geographic areas with a name can be contentious and open to debate. Names are expected to be broadly representative of the characteristics of the entire group. However, it is important to recognize that names attached to clusters are only indicative of the predominant features of the geographic areas that make up the clusters.

Some principles were considered while naming the clusters that emerged from the analysis. Names should not be offensive in any way. In a multiethnic and multireligious country like Nigeria, an offensive label can quickly provoke problematic interpretations. Careful consideration was therefore given to chosen labels to ensure that they do not stigmatize any section of the Nigerian public. Religious and ethnic tags were also avoided as much as possible. Figure 8.5 shows the labels given to different clusters. The six clusters created in the first hierarchy are called supergroups. The 23 clusters at the second level of the hierarchy are called groups, while the third hierarchy comprises 57 clusters called subgroups.

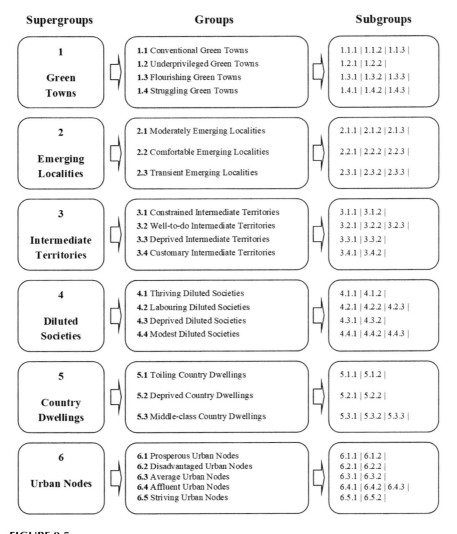

FIGURE 8.5

Hierarchical structure and labels of Nigerian clusters.

In a sense, the labels given to clusters in Figure 8.5 attempt to reflect some of the distinguishing characteristics of the areas that make up those clusters. Distinguishing variables have been identified by looking at the resultant z-scores of the final cluster centers. The mean for each cluster was calculated and variables with values higher than the mean stood out as distinguishing variables.

Radial plots as displayed in Figure 8.6 have been created for all supergroups, groups, and subgroups. In addition to the charts, pen portraits that are textual descriptions of the supergroups and groups have also been developed. Pen portraits that summarize the prevalent characteristics of each cluster have the benefit of elucidating (in qualitative terms) some of the information inherent in complex quantitative analysis. Figure 8.7 shows the spatial distribution of the 23 groups mapped at LGA scale.

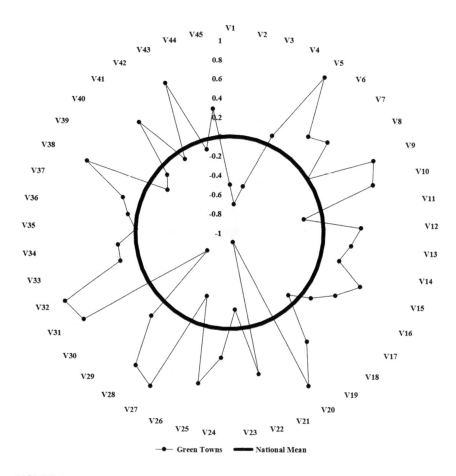

FIGURE 8.6
Radial plot of green towns (Author's elaboration).

V1	Use of agricultural inputs	V24	Access to safe sources of water
V2	Owns cattle	V25	No toilet facility
V3	Age 0–14 years	V26	Safe toilet sanitation
V4	Age 15–59 years	V27	Nonwood fuel for cooking
V5	Age 60 years and over	V28	Lighting energy-mains electricity
V6	Never married	V29	Access to water supply
V7	At least one pensioner	V30	Own a house
V8	Population density	V31	Renting a house
V9	Separated couples	V32	Free accommodation
V10	Completed secondary education	V33	Built with burnt bricks
V11	Head of household uneducated	V34	Single room
V12	Adult literacy rate	V35	Duplex
V13	Access to primary school	V36	Built with cement/mud brick
V14	Economically active population	V37	Vehicle ownership
V15	Self-employed	V38	Motorcycle ownership
V16	Employment in the transport sector	V39	Size of land over 6 hectares
V17	Employment in agriculture	V40	Improved security
V18	Taking anti-malaria measures	V41	Access to transportation
V19	Access to health	V42	Difficulty in meeting basic needs
V20	Household size 1–2 persons	V43	Children living with single parents
V21	Mean household size	V44	Early marriage and teenage pregnancy
V22	Ownership of mobile phone	V45	Vaccinated
V23	Ownership of personal computer		

FIGURE 8.6
(Cont.)

8.5 Conclusion

The discussions in this chapter have explained how a geodemographic classification was developed for Nigerian LGAs. No two classifications are developed in the same way. The discussions in this chapter are essentially a set of justifications for the decisions made throughout the process. The methods used for the analysis center around unsupervised machine learning and spatial analysis. It is fair to say that the results generated from this analysis indicate that developing countries can also benefit from geodemographic methods despite difficulties surrounding access to data. An important lesson reported in this chapter is that the process of choosing input variables should not be a one-size-fits-all approach. By itself, however, the area classification discussed here may not answer all the questions without

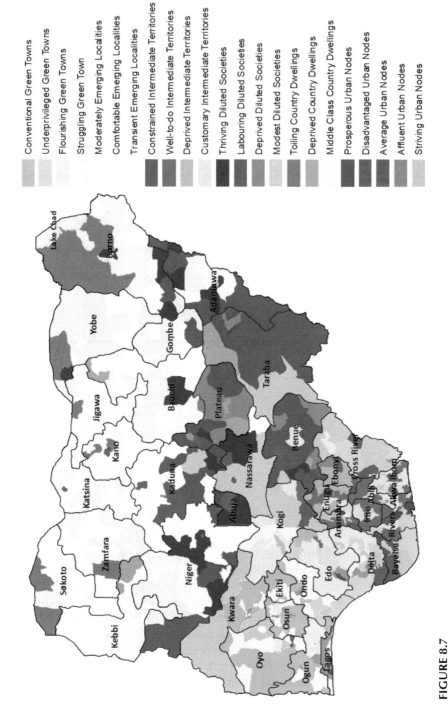

FIGURE 8.7

Map of the 23 groups (Author's elaboration previously published in Ojo, A., Vickers, D. and Ballas, D. (2012). The Segmentation of Local Government Areas: Creating a New Geography of Nigeria. *Applied Spatial Analysis and Policy*, 5(1), 25–49).

linkage to additional ancillary data and further in-depth spatial analysis as demonstrated in Chapter 10.

References

Adesugba, M.A. and Mavrotas, G. (2016). *Youth Employment, Agricultural Transformation, and Rural Labor Dynamics in Nigeria*. IFPRI Discussion Paper 1579. International Food Policy Research Institute (IFPRI), Washington, D.C.

Ballas, D., Rossiter, D., Thomas, B., Clarke, G. and Dorling, D. (2005). *Geography Matters: Simulating the Local Impacts of National Social Policies*. York: Joseph Rowntree Foundation.

Bloom, D.E. and Sachs, J.D. (1998). Geography, Demography and Economic Growth in Africa. *Brookings Papers on Economic Activity*, (2), 207–295.

Everitt, B.S., Landau, S. and Leese, M. (2001). *Cluster Analysis*. London: Arnold.

Federal Government of Nigeria (1987). *Nomadic Education Blueprint*. Lagos: Government Printers.

Gordon, A.A. (2003). *Nigeria's Diverse Peoples: A Reference Source Book*. Santa Barbara, CA: ABC-CLIO.

Gregory, I.N. (2000). *An Evaluation of the Accuracy of the Areal Interpolation of Data for the Analysis of Long-Term Change in England and Wales*. Available at: www.geocomputation.org/2000/GC045/Gc045.htm. Accessed on: 15 February 2020.

Hagen-Zanker, J. and Holmes, R. (2012). *Social protection in Nigeria: Synthesis Report*. London: Overseas Development Institute.

Harris, R., Sleight, P. and Webber, R. (2005). *Geodemographics, GIS and Neighborhood Targeting*. London: Wiley.

House, J.S., Kessler, A.R. and Herzog, R. (1990). Age, Socioeconomic Status and Health. *The Milbank Quarterly*, 68(3), 383–411.

Iro, I. (2006). *Nomadic Education and Education for Nomadic Fulani*. Available at: www.gamji.com/fulani7.htm (Accessed: 16 February 2020).

Jagun, A., Heeks, R. and Whalley, J. (2008). The Impact of Mobile Telephony on Developing Country Micro-Enterprise: A Nigerian Case Study. *Information Technologies and International Development*, 4(4), 47–65.

Morgan, L.A. (1991). *After Marriage Ends: Economic Consequences for Midlife Women*. Newbury Park, CA: Sage Publications.

NBS (2006). *Core Welfare Indicators Questionnaire Survey: Final Statistical Report*. Abuja: National Bureau of Statistics.

Nwankpa, N. (2017). Sustainable Agricultural Development in Nigeria: A Way Out of Hunger and Poverty. *European Journal of Sustainable Development*, 6(4), 175–184.

Ogunbodede, E.F. (2006). Developing Geospatial Information for Poverty Reduction: Lessons and Challenges from Nigeria's 2006 Census. *GSDI-9 Conference Proceedings*. Santiago, Chile, 6–10 November.

Ojo, A. and Ezepue, P.O. (2012). Modeling and Visualizing the Geodemography of Poverty and Wealth across Nigerian Local Government Areas. *The Social Sciences*, 7(1), 145–158.

Ojo, A. and Ojewale, O. (2019). *Urbanisation and Crime in Nigeria*. Cham: Palgrave Macmillan.

Ojo, A., Vickers, D. and Ballas, D. (2012). The Segmentation of Local Government Areas: Creating a New Geography of Nigeria. *Applied Spatial Analysis and Policy*, 5(1), 25–49.

Okonofua, F. (2010). Reducing Maternal Mortality in Nigeria: An Approach through Policy Research and Capacity Building. *African Journal of Reproductive Health*, 14(3), 9–14.

Palmer, R., Wedgwood, R., Hayman, R., King, K. and Thin, N. (2007). *Educating Out of Poverty? A Synthesis Report on Ghana, India, Kenya, Rwanda, Tanzania and South Africa*. Edinburgh: Centre of African Studies, University of Edinburgh.

Turnwait, M.O. and Odeyemi, M.A. (2017). Nigeria's Population Policies: Issues, Challenges and Prospects. *Ibadan Journal of the Social Sciences*, 15(1), 104–114.

UN (2019). *World Population Prospects 2019: Online Edition. Rev. 1*. New York, NY: United Nations Department of Economic and Social Affairs, Population Division.

Vickers, D.W. (2006). *Multi-Level Integrated Classifications Based on the 2001 Census*. PhD thesis. School of Geography, University of Leeds, United Kingdom.

Vickers, D. and Rees, P. (2006). Introducing the Area Classification of Output Areas. *Population Trends*, 125, 15–29.

WHO (2019). *World Malaria Report 2019*. Geneva: World Health Organization.

9

Combining Continuous and Categorical Data to Segment Philippines Barangays

9.1 Administrative Divisions of the Philippines

The Philippines is located on the eastern side of the Southeast Asia. Population estimates for 2020 derived from the 2019 UN World Population Prospects indicate that the Philippines is home to 109 million people (UN, 2019). The Philippine archipelago is consisted of more than 7,000 islands, with a total land area of approximately 300,000 square kilometers (NSO, 2008). About 1% (1,830 square kilometers) of this area is water. According to Coursey (2008), the group of islands is naturally divided into four segments. In the northern area, there is Luzon, which is the largest amongst the four groups of islands. Luzon comprises areas like Palawan, Mindoro, Masbate, and Marinduque. The Visayas is another cluster of islands south of Luzon. This group is located in the central area of the country. Some of the important islands within the Visayas include Samar, Leyte, Bohol, Cebu, Negros, and Panay. The second largest group of islands is located in the southernmost part of the country. This group is called Mindanao. It comprises Camiguin, Basilan, and a host of other islands including the Sulu archipelago. The fourth group of islands is located in the southwestern division and is called Palawan. This group comprises a cluster of islands extending in a narrow manner and connecting islands in the Luzon group with the Visayas.

The administrative geography of the Philippines is technically consisted of four hierarchies, which include regions, provinces, cities, municipalities, and *barangays*. Over the years, the number of spatial units that make up each type of administrative division has changed. Table 9.1 shows the number of spatial units that made up each administrative division as of 2019.

Regions provide government departments with bases to establish their offices. Figure 9.1 shows a map of the regions of the Philippines. According to the last census conducted in 2015, the average population of each region was 5.9 million people. Calabarzon region accounts for 14% (the highest) of the national population, followed by the National Capital Region that accounts for 13%. The other double-digit percentage was recorded in Central Luzon,

TABLE 9.1

Administrative divisions of the Philippines

Administrative Division	Number of Spatial Units (2019)
Regions	17
Provinces	81
Municipalities	1,489
Barangays	42,045

Source: Author's compilation

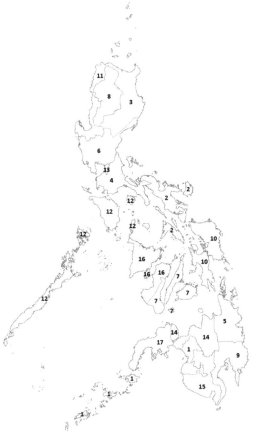

1 - Autonomous Region

2 - Bicol

3 - Cagayan Valley

4 - Calabarzon

5 - Caraga

6 - Central Luzon

7 - Central Visayas

8 - Cordillera Administrative Region

9 - Davao

10 - Eastern Visayas

11 - Ilocos

12 - Mimaropa

13 - National Capital Region

14 - Northern Mindanao

15 - Soccsksargen

16 - Western Visayas

17 - Zamboanga Peninsula

FIGURE 9.1

Regions of the Philippines.

which accounts for 11% of the national population. Cordillera Administrative Region accounts for only 2% of the national population, which is the least amongst all the regions. As of 2015, the regional population range was 12.7 million people.

Provincial spatial units are embedded within regions. Each province has a level of autonomy and provinces are governed by two arms of government that have executive and legislative powers (Woods, 2006). Each province is headed by a governor. Local level governance is provided at municipality level in the Philippines. Municipalities are subdivisions of provinces. Each municipality is governed by a mayor who presides over executive issues (Woods, 2006). *Barangays* make up the most granular level of geography. A *barangay* is a village or a very small district. They are small communities expected to provide residential accommodation to populations of approximately 500 households (NSO, 2008). *Barangays* with projected households not exceeding 500 are considered as enumeration areas. Social ties are generally tightly knitted at the *barangay* administrative division (Woods, 2006).

9.2 Social Aspects

The current inhabitants of the Philippines are thought to have historical links with migrants from Indonesia, China, and Malaysia (Worcester, 2005; Coursey, 2008). For about 350 years from 1556, the country was under the Spanish Colonial rule that ended in 1896 (Woods, 2006). However, the collapse of the Spanish did not lead to its independence. A new era of colonial control by the United States commenced in 1898. Philippines gained independence from the United States on July 4, 1946 (Woods, 2006; Coursey, 2008) but had its economy predominantly linked to the economy of the United States.

The Philippines has more than 170 languages (Gordon, 2005). Two of these (English and Filipino/Pilipino) are the national/official languages of the country. Although the Philippines is characterized by a rich and diverse culture, three ethnolinguistic groups constitute the majority share (50%) of the population. The Tagalogs make up 28% of the population, the Cebuano ethnolinguistic groups make up 13%, and the Ilocanos account for 9% of the population of the country. Other important groups include the Bisaya and Hiligaynon that constitute 8% each, the Bikol that account for 6% of the population, and the Waray that make up 3% of the population. There are several other minority population groups that constitute 24% of the population (NSO, 2008).

Many of the indigenous traditions of the Philippines have been influenced by American and Hispanic cultures largely due to the long periods of colonialism. Geography is correlated with the cultural distributions of Filipinos. The Ilocanos concentrate in the north, the Tagalogs are mostly found in the central plains of the country, while the Visayans are clustered in the central Islands (Worcester, 2005; Woods, 2006). Most of the other groups are scattered across the archipelago.

Since the country is an island archipelago, several aspects of its existence are dominated by aquatic life. About 1.5 million square kilometers of territorial water serves about 59 lakes and 132 major rivers, which provide rich fishing resources. The country has a thriving tourism industry. Some of the key tourist destinations include Cebu, Bohol, Palawan, and Boracay.

9.3 Data Source – The 2000 Census of the Philippines

Like many other developing countries, spatial analytical techniques for decision-making in the Philippines are still evolving. Consequently, researchers face some bottlenecks when trying to conduct granular level spatial analysis. Data access impediments are largely due to technical challenges, lack of current data, and bureaucratic barriers (Saxena, 2018).

On May 1, 2000, the first census of the present millennium was conducted in the Philippines. The census is traditionally the most comprehensive body of information about population sociodemographics and housing. Data from the 2000 Census has been used to develop a geodemographic classification for the Philippines at the *barangay* level. The sheer volume of data collected during censuses makes data processing an intimidating task. Consequently, data from the Philippines 2000 Census were released in phases. From the 2000 Census, the estimated population of the country was about 76.5 million people (NSO 2010). Census data sets were derived from two sources. The first source was the National Statistics Office (NSO) in Manila, which provided GIS digital boundaries as well as census data for multiple geographic levels. Much of the data secured from the NSO were categorical variables describing the presence or absence of public facilities and services at *barangay* scale. Data were not made available for 14 *barangays*. Further correspondence with the NSO staff revealed that these *barangays* either were evacuated during the conduct of the census due to armed clashes between rebel forces and government troops, or had been evacuated due to volcanic eruption at Mount Pinatubo. These 14 *barangays* were excluded from the analysis.

The second source of data was derived from the Integrated Public Use Microdata Series-International (IPUMS-International). IPUMS is based at the University of Minnesota in Minneapolis, USA. The data from IPUMS comprised a large database of nearly 7.5 million individual records forming 10% of the census at municipality level.

9.4 Evaluating Categorical and Continuous Data for Small Areas

Municipality- and *barangay*-level data sets were combined to develop the area classification discussed in this chapter at the *barangay* scale. Webber (2004) found that by incorporating data from multiple levels of geography, the discriminatory power of small area geodemographic classifications could be enhanced. In addition to providing summary characteristics of residents of a geographic jurisdiction, small area classifications also try to explain the influence of immediate and adjoining communities on people. Webber (2004) described this influence as *neighborhood effects*.

A total of 435 variables derived from the two data sources. For ease of understanding, all the variables were assembled under ten themes. The ten data themes observed within the data set include:

- Demographics
- Education
- Employment
- Health
- Household infrastructure
- Housing
- Religion
- Services
- Socioeconomics
- Transportation

A total of 161 (37%) variables were from the NSO, while the remaining 63% were from the IPUMS. Of the 435 variables, 5% were categorical while 95% were continuous variables. Each variable was considered as a potential candidate for inclusion in the analysis. Preprocessing of initial variables was done in three phases. First, all 23 categorical variables were analyzed together because the statistical tests deployed on categorical data were slightly different from those deployed on continuous data. Second, all continuous data sets were evaluated on a theme-by-theme basis. Third, all the continuous variables that were chosen after the theme-by-theme data reduction were evaluated together. At every stage in the quantitative analysis, the importance of each variable for key policy issues was given due consideration.

9.4.1 Assessing Geographic Dispersion

The techniques used to preprocess categorical data are generally different from those used to prepare continuous data. For adequate distinctions between areas, input variables need to demonstrate reasonable levels of geographic variation. The standard deviation is a useful statistic for measuring the geographic dispersion of continuous data. Standard deviation (σ), defined as the square root of the variance (Crawshaw and Chambers, 2001), provides a measure of the magnitude by which values tend to depart from the mean.

For the categorical data sets, a different method was adapted to calculate the mean and convert the variables into numbers. This was done by deriving the weighted average of all probable values that each variable can take on. The weights used in calculating the average correspond to probabilities. Crawshaw and Chambers (2001) provided detailed explanation on how an expectation or expected mean can be calculated for discrete categorical distributions using probability theory. Table 9.2 shows the categorical variables arranged in ascending order of their variation across geographic spaces.

9.4.2 Measuring Association

Pearson's chi-square test has been found to be a useful statistic for analyzing categorical data (Johnson et al., 1994; Maydeu-Olivares and Joe, 2014). The chi-square test is often used to test the goodness of fit between a theoretical and a frequency distribution. It is also used to investigate dependency between variables. However, there are two major shortcomings of the chi-square test: first, it does not give an indication of the direction of the association (i.e. positive or negative) between variables. Second, the chi-square test is sensitive to sample size. A more useful measure of the strength of association among categorical variables is to calculate the proportional difference between concordant (P) and discordant (Q) pairs (Agresti, 2018) and derive a measure called the gamma, as defined in Equation 9.1.

$$\gamma = \frac{P-Q}{P+Q} \tag{9.1}$$

where:

γ represents the gamma statistic value of the variable

P represents the total number of pairs that rank the same (concordant pairs)

Q represents the number of pairs that do not rank the same (discordant pairs).

Gamma (γ) can take values ranging from -1 to $+1$. A value closer to $+1$ indicates a positive (concordant) association, while a value closer to -1 indicates a negative (discordant) relationship. To derive the gamma, a

TABLE 9.2
Standard deviation values for categorical variables

Variables	Expectation $E(x)$ or μ	μ^2	$E\!\left(x^2\right)$	$Var(x)$	$SD(\bar{A})$
Does the *barangay* have a street pattern or network of streets of at least 3 street roads?	1.50	2.24	2.49	0.26	0.51
Is *barangay* part of town city proper or poblacion?	1.55	2.41	2.66	0.25	0.50
Is there a community waterworks system in the *barangay*?	1.52	2.32	2.56	0.24	0.49
Is there a puericulture center/health center in the *barangay*?	1.39	1.93	2.16	0.23	0.48
Is there a public plaza or park in the *barangay*?	1.64	2.70	2.92	0.22	0.47
Is there an elementary school in the *barangay*?	1.27	1.60	1.79	0.19	0.43
Is there a telephone in the *barangay*?	1.73	3.00	3.18	0.19	0.43
Is there electric power in the *barangay*?	1.24	1.54	1.71	0.17	0.41
Is the *barangay* accessible to the national highway?	1.22	1.48	1.65	0.17	0.41
Is there a cemetery in the *barangay*?	1.79	3.19	3.35	0.16	0.40
Is there a postal service in the *barangay*?	1.81	3.28	3.42	0.14	0.38
Is there a market or building with trading activities in the *barangay*?	1.82	3.33	3.47	0.14	0.38
Is there a church/chapel or mosque in the *barangay*?	1.18	1.38	1.52	0.14	0.37
Is there a high school in the *barangay*?	1.83	3.35	3.48	0.13	0.37
Is there a newspaper circulation in the *barangay*?	1.84	3.37	3.50	0.13	0.36
Is there a *barangay* hall in the *barangay*?	1.13	1.27	1.37	0.10	0.32
Is there a housing project (government or private) in the *barangay*?	1.91	3.63	3.71	0.08	0.27
Is there a town/city hall or provincial capital in the *barangay*?	1.93	3.73	3.79	0.06	0.25
Is there a telegraph in the *barangay*?	1.93	3.74	3.79	0.05	0.22
Is there a hospital in the *barangay*?	1.95	3.81	3.85	0.03	0.19
Is there a college/university in the *barangay*?	1.97	3.87	3.89	0.03	0.16
Is there a public library in the *barangay*?	1.97	3.87	3.89	0.02	0.14

Source: Author's calculation previously published in Ojo, A., Vickers, D. and Ballas, D. (2013). Creating a Small Scale Area Classification for Understanding the Economic, Social and Housing Characteristics of Small Geographical Areas in the Philippines. *Regional Science Policy and Practice*, 5(1), 1–24.

crosstab query was run on all categorical variables to derive two-by-two matrices for each pair. Each of the two-by-two matrices was ordered by "yes" and "no" responses as shown in Table 9.3.

From the data in Table 9.3, 50.4% (21,141) of all *barangays* have both a health centre and an elementary school, while 15.3% (6,347) have none. The concordant (P) for the table was derived by multiplying 21,141 by 6,347, while the discordant (Q) was derived by multiplying 4,464 by 9,498. The output of the ratio of the difference and the sum of these two statistics is the gamma. The strength and direction of the association between the different pairs of

TABLE 9.3

Cross-tabulation of two categorical variables

Variables		Is There an Elementary School in the *Barangay*?	
		Yes	No
Is there a health center in the	Yes	21141	4464
barangay?	No	9498	6347

Source: Author's calculation.

categorical variables are presented in the matrix in Table 9.4. Generally, there is a weak positive association between the data sets. The shaded cells in Table 9.4 indicate variables with high levels of positive correlation ($\gamma = +0.7$ and above). For instance, there is a very strong positive relationship between the presence or absence of a telegraph and postal services. This positive relationship indicates that the presence of one service in a *barangay* may suggest the presence of the other and the absence of one service may suggest the absence of the other. This may also suggest that users of telegraph services may prefer to live in a neighbourhood where there is also a postal service. Including multiple variables with strong levels of association can result in redundant information that can cloud other underlying characteristics. The product–moment correlation coefficient was used to explore the strength and direction of relationships among continuous variables.

9.4.3 The Analysis of Principal Components

PCA was used to determine those variables that are likely to have the greatest influence on the classification. PCA was conducted for both categorical and continuous variables in order to determine those variables that are likely to power the data set. Although the components loading matrices of the first three components were examined, there had to be trade-off on which a component would be used to determine the variables that power the data set. The first principal component accounts for the greatest possible proportion of the variance of the variable set, while the second component accounts for the maximum remaining variance. Essentially later components explain less of the variation within the data (Jolliffe, 2002). The first component was therefore used to determine those variables that power the data set.

Table 9.5 shows the results of PCA conducted for categorical variables. The presence or absence of a postal and a telephone service are the two categorical indicators that are likely to power the data set, with 42% of the variance existing within these variables explained by component 1. Among continuous variables, population without access to piped water had the highest loading factor for component 1.

TABLE 9.4

Gamma statistic matrix for categorical variables

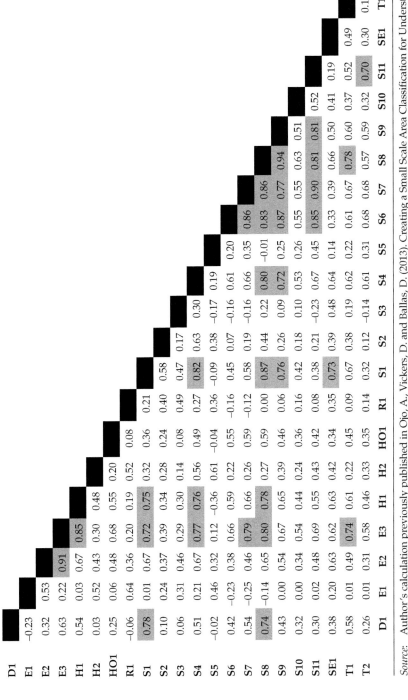

	D1	E1	E2	E3	H1	H2	HO1	R1	S1	S2	S3	S4	S5	S6	S7	S8	S9	S10	S11	SE1	T1	T2
D1																						
E1	−0.23																					
E2	0.32	0.53																				
E3	0.63	0.22	0.91																			
H1	0.54	0.03	0.67	0.85																		
H2	0.03	0.52	0.43	0.30	0.48																	
HO1	0.25	0.06	0.48	0.68	0.55	0.20																
R1	−0.06	0.64	0.36	0.20	0.19	0.52	0.08															
S1	0.78	0.01	0.67	0.72	0.75	0.32	0.36	0.21														
S2	0.10	0.24	0.37	0.39	0.34	0.28	0.24	0.40	0.58													
S3	0.06	0.31	0.46	0.29	0.30	0.14	0.08	0.49	0.47	0.17												
S4	0.51	0.21	0.67	0.77	0.76	0.56	0.49	0.27	0.82	0.63	0.30											
S5	−0.02	0.46	0.32	0.12	−0.36	0.61	−0.04	0.36	−0.09	0.38	−0.17	0.19										
S6	0.42	−0.23	0.38	0.66	0.59	0.22	0.55	−0.16	0.45	0.07	−0.16	0.61	0.20									
S7	0.54	−0.25	0.46	0.79	0.66	0.26	0.59	−0.12	0.58	0.19	−0.16	0.66	0.35	0.86								
S8	0.74	−0.14	0.65	0.80	0.78	0.27	0.59	0.00	0.87	0.44	0.22	0.80	−0.01	0.83	0.86							
S9	0.43	0.00	0.54	0.67	0.65	0.39	0.46	0.06	0.76	0.26	0.09	0.72	0.25	0.87	0.77	0.94						
S10	0.32	0.00	0.34	0.54	0.44	0.24	0.36	0.16	0.42	0.18	0.10	0.53	0.26	0.55	0.55	0.63	0.51					
S11	0.30	0.02	0.48	0.69	0.55	0.43	0.42	0.08	0.38	0.21	−0.23	0.67	0.45	0.85	0.90	0.81	0.81	0.52				
SE1	0.38	0.20	0.63	0.62	0.63	0.42	0.34	0.35	0.73	0.39	0.48	0.64	0.14	0.33	0.39	0.66	0.50	0.41	0.19			
T1	0.58	0.01	0.49	0.74	0.61	0.22	0.45	0.09	0.67	0.38	0.19	0.62	0.22	0.61	0.67	0.78	0.60	0.37	0.52	0.49		
T2	0.26	0.01	0.31	0.58	0.46	0.33	0.35	0.14	0.32	0.12	−0.14	0.61	0.31	0.68	0.68	0.57	0.59	0.32	0.70	0.30	0.17	
	D1	E1	E2	E3	H1	H2	HO1	R1	S1	S2	S3	S4	S5	S6	S7	S8	S9	S10	S11	SE1	T1	T2

Source: Author's calculation previously published in Ojo, A., Vickers, D. and Ballas, D. (2013). Creating a Small Scale Area Classification for Understanding the Economic, Social and Housing Characteristics of Small Geographical Areas in the Philippines. *Regional Science Policy and Practice*, 5(1), 1–24.

(*continued*)

TABLE 9.4
(Cont.)

D1:	Is *barangay* part of town city proper or poblacion?	S4:	Is there a public library in the *barangay?*
E1:	Is there an elementary school in the *barangay?*	S5:	Is there a *barangay* hall in the *barangay?*
E2:	Is there a high school in the *barangay?*	S6:	Is there a newspaper circulation in the *barangay?*
E3:	Is there a college/university in the *barangay?*	S7:	Is there a telephone in the *barangay?*
H1:	Is there a hospital in the *barangay?*	S8:	Is there a telegraph in the *barangay?*
H2:	Is there a health center in the *barangay?*	S9:	Is there a postal service in the *barangay?*
HO1:	Is there a housing project in the *barangay?*	S10:	Is there a community waterworks system in the *barangay?*
R1:	Is there a church/chapel or mosque in the *barangay?*	S11:	Is there electric power in the *barangay?*
S1:	Is there a town/city hall or provincial capital in the *barangay?*	SE1:	Is there a market or building with trading activities in the *barangay?*
S2:	Is there a public plaza or park in the *barangay?*	T1:	Does the *barangay* have a street pattern of at least three street roads?
S3:	Is there a cemetery in the *barangay?*	T2:	Is the *barangay* accessible to the national highway?

TABLE 9.5
Top ten loadings of the first principal component of categorical variables

Categorical Variable	Loading
Is there a postal service in the *barangay*?	0.65
Is there a telephone in the *barangay*?	0.65
Is there a telegraph in the *barangay*?	0.62
Is there a newspaper circulation in the *barangay*?	0.58
Does the *barangay* have a street pattern of at least three street roads?	0.54
Is there a high school in the *barangay*?	0.49
Is there a town/city hall or provincial capital in the *barangay*?	0.48
Is there a market or building with trading activities in the *barangay*?	0.44
Is there a college/university in the *barangay*?	0.43
Is there electric power in the *barangay*?	0.43

Source: Author's calculation previously published in Ojo, A., Vickers, D. and Ballas, D. (2013). Creating a Small Scale Area Classification for Understanding the Economic, Social and Housing Characteristics of Small Geographical Areas in the Philippines. *Regional Science Policy and Practice*, 5(1), 1–24.

TABLE 9.6
The final list of variables

Variable	Data Theme	Variable	Data Theme
Part of city or poblacion	Demographics	Electricity supplied	Household infrastructure
Population density	Demographics	No piped water	Household infrastructure
No married couples	Demographics	Cooking energy: liquid fuels	Household infrastructure
Two or more married couples	Demographics	Cooking energy: wood	Household infrastructure
Stepchild of head	Demographics	No telephone	Household infrastructure
Age 0–5 years	Demographics	Television	Household infrastructure
Age 6–10 years	Demographics	Water closet toilet	Household infrastructure
Age 11–20 years	Demographics	Latrine	Household infrastructure
Age 21–44 years	Demographics	Dwelling owned	Housing
Age 45–64 years	Demographics	Members squatting	Housing
Age 65+ years	Demographics	Free occupancy	Housing
Widowed	Demographics	Land is occupied with consent	Housing
Speaks Filipino	Demographics	Built with bamboo	Housing
Elementary school	Education	Built with bricks, stone, or concrete	Housing
High school	Education	Built with mixed materials	Housing
Less than primary education	Education	Iron and concrete roofing	Housing
Postsecondary technical education	Education	Buddhist	Religion

(continued)

TABLE 9.6
(Cont.)

Variable	Data Theme	Variable	Data Theme
Professionals and senior officials	Employment	Muslim	Religion
Agricultural and fishery workers	Employment	Roman Catholic	Religion
Transportation and communications employment	Employment	Other Christians	Religion
Public administration and defense employment	Employment	Town hall	Services
Education employment	Employment	Public plaza or park	Services
Health and social work employment	Employment	Cemetery	Services
Overseas worker	Employment	*Barangay* hall	Services
Real estate and business establishments	Employment	Telegraph service	Services
Private establishments	Employment	Postal service	Services
Cooperative establishments	Employment	Waterworks system	Services
Small establishments	Employment	Method of wastewater disposal	Services
Media establishments	Employment	Market	Socioeconomic
Retail trade establishments	Employment	Manufacturing establishments	Socioeconomic
Banking institutions	Employment	Auto repair shops	Socioeconomic
Recreational establishments	Employment	Restaurants and personal services	Socioeconomic
Hospital	Health	At least three street roads	Transportation
Health center	Health	Accessibility to the highway	Transportation
Disabled	Health		

Source: Author's compilation previously published in Ojo, A., Vickers, D. and Ballas, D. (2013). Creating a Small Scale Area Classification for Understanding the Economic, Social and Housing Characteristics of Small Geographical Areas in the Philippines. *Regional Science Policy and Practice*, 5(1), 1–24.

9.4.4 The Final List of Variables

The preprocessing of data was a lengthy and rigorous process that required making complex decisions. In addition to the statistical tests so far discussed, the skew of different variables was also evaluated. The judgments made in the selection process were carefully considered and are by no means finite. Table 9.6 shows the list of the 69 variables selected for further analysis and inclusion in the subsequent clustering algorithm.

The initial list comprised a total of 435 variables, with the demographic and employment themes recording the largest shares of 31% and 25%, respectively. At the end of the first phase of the data preprocessing, 81% of the initial variables were excluded, which reduced the number of variables to

81. At this stage, the proportion of all variables accounted for by the demographic and employment themes had reduced to 44%. The second phase of data preprocessing focused on merged variables and inter-theme analysis. Another 15% of the variables were rejected during this phase. Thus, the final list comprised 69 variables. The employment and demographic themes accounted for the largest proportions of 22% and 19%, respectively. Three themes – household infrastructure, housing, and services – each accounted for 12%. The education, religion, and socioeconomic themes accounted for 6% each. The health theme had a share of 4%, while the transportation theme accounted for 3% of all variables.

9.5 Grouping *Barangays* Based on Demographic and Housing Characteristics

Prior to proceeding with clustering, the different measurement scales of the input variables were considered. To rescale the data set, variables were converted into standard normal variate scores. A logarithmic transformation was applied to reduce the effect of skew within the data set. This method was used because of its ability to cope well with positive skew.

Once the data set had been prepared, a choice of clustering algorithm was to be made. The size of the data required an algorithm that could handle the volume of data. The other issue to consider was the fact that the database was a combination of categorical and continuous data sets. Most clustering algorithms work well with continuous data sets. An algorithm specifically designed to handle a combination of categorical and continuous data sets is the two-step cluster analysis procedure (Banerjee et al., 2014). This procedure gives the best results if categorical data sets appear to have a multinomial distribution and continuous variables display a normal distribution.

During the first step of the clustering procedure, pre-clusters are formed using a distance measure. The pre-clustering process does not require a prespecification of the desired number of clusters. Every object is considered in relation to already formed clusters, and based on the distance measure, it is decided whether an object should start a new pre-cluster or be assigned to an already formed cluster (Hellerstein and Stonebraker, 2005).

Once the pre-clusters have been formed, each of the clusters is considered as a single object. The second step of the procedure deploys a hierarchical algorithm on the pre-clusters (Hellerstein and Stonebraker, 2005). During this step, attention is given to the number of pre-clusters formed. A large number of pre-clusters result in better solutions (Hellerstein and Stonebraker, 2005) but demand more computational power, which in turn slows down the algorithm. All clustering algorithms group cases based on similarity or dissimilarity. The similarity of cases within a taxonomic space is measured by deriving a statistical quantification of distance (Everitt et al., 2001). Generally, similar cases have a closer distance. Agresti (2013) has recommended a likelihood ratio test as a reliable test for significance when dealing with categorical data sets. The log-likelihood ratio test was used to evaluate similarity within the data sets. It is based on probabilities and compares the maximized likelihoods of a null hypothesis with an alternative one. The larger the value of the statistic, the less the within-cluster variation and the more compact cases are.

9.5.1 Clustering Criterion and Number of Clusters

In order to create hierarchical classification system using the two-step clustering method, an acceptable number of clusters had to be derived at the top level of the hierarchy. One of the approaches suggested by Everitt et al. (2001) is to evaluate the clustering criterion against the number of clusters. The clustering criterion selected for the exercise was Schwarz's Bayesian information criterion (BIC), which is given in the notation in Equation 9.2.

$$BIC = -2.\ln L + k\ln(n) \qquad (9.2)$$

where:

- n represents the number of data points in the observed data
- k represents the number of free parameters to be estimated
- L represents the maximized value of the likelihood function for the estimated model.

The change in BIC is basically the difference between log-likelihood ratio statistic and the clustering parameters (Akaike et al., 1998). A perceived ideal solution would be the point at which there is an abrupt change in the BIC (Larose, 2006; Banerjee et al., 2014). As shown in Figure 9.2, this point is located at the seven-cluster solution after running the analysis for up to 50 clusters. After carefully deciding on the number of clusters at the top level of the hierarchy, other levels of the hierarchy were created.

9.6 Profiles, Pen Portraits, and Mapping

Labeling of clusters in a small area classification is an important part of the exercise because the process of naming brings the area classification into life. It was decided that a two-word label may not sufficiently summarize the multivariate characteristics of clusters. Additionally, the labels assigned to

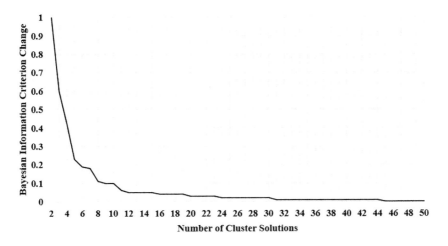

FIGURE 9.2

Change in the average distance of cases to cluster center (Author's elaboration previously published in Ojo, A., Vickers, D. and Ballas, D. (2013). Creating a Small Scale Area Classification for Understanding the Economic, Social and Housing Characteristics of Small Geographical Areas in the Philippines. *Regional Science Policy and Practice*, 5(1), 1–24).

clusters were informed by conducting in-depth analysis of some statistical characteristics of the clusters.

9.6.1 Between-Cluster and Within-Cluster Variation of Variables

An understanding of the distribution of variables within each cluster can be very helpful for labeling process. Since the area classification of the Philippines combines categorical and continuous variables, the composition of clusters was evaluated using different methods for both types of variables. Categorical variables were assessed by looking at the cross-tabulations of the percentage of the variables within each cluster. Figure 9.3 shows the pattern of variation of a categorical variable (*the presence or absence of at least three street roads*) within each of the seven clusters.

The overall distribution of the variable across all *barangays* is balanced. Cluster 1 shows that street roads are absent in many *barangays* that make up the cluster . In clusters 2, 3, and 4, there is also a higher degree of absence of street roads. Cluster 5 displays a disproportionately high presence of three street roads in comparison to all other clusters. Both clusters 6 and 7 also have a much larger presence of street roads. This type of categorical variable analysis helps us to identify those characteristics that are dominant within each cluster. Dominant characteristics can be assembled to help guide the labeling of clusters.

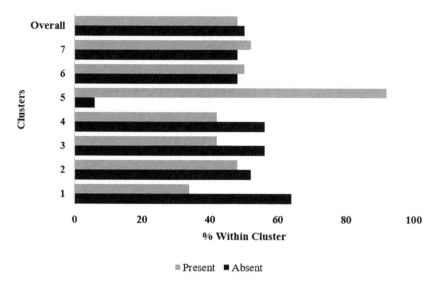

FIGURE 9.3
Within-cluster variation for the variable *Three Street Roads* (Author's elaboration).

9.6.2 Statistical Significance of Variables

Another useful way of using statistics to guide the labeling process is to test for the statistical significance of variables within each cluster. A variable may show a strong presence within a cluster as illustrated in Section 9.6.1, but it may not be statistically significant. Similarly, a variable may show a weak presence within a cluster, but it may be statistically significant. The statistical significance of categorical variables by clusters was analyzed by comparing the observed distribution of each variable with the expected distribution using the chi-square statistic. Figure 9.4 shows the statistical significance of categorical variables in the formation of cluster 2.

From Figure 9.4, it is evident that the presence or absence of a telegraph service is the most significant categorical variable for cluster 2, while the presence or absence of a cemetery is the least important variable.

9.6.3 Cluster Labels and Mapping

The results discussed in Sections 9.6.1 and 9.6.2 reinforce the importance of using statistical approaches for understanding the characteristics of different clusters and those features that make them unique. The knowledge garnered from the analysis also contributed to the naming procedure. Cluster names and hierarchical structure are shown in Figure 9.5.

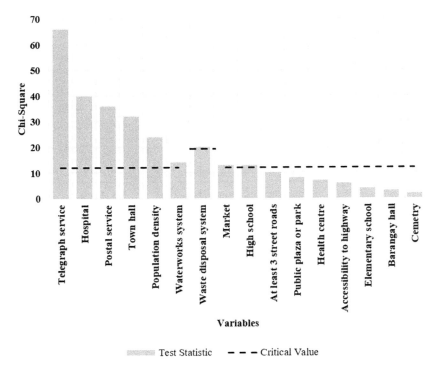

FIGURE 9.4
Significance of categorical variables for cluster 2 (Author's elaboration).

It is important for the number of *barangays* and population that are assigned to each cluster to be reasonably balanced. This will ensure that the solution is robust and suitable for use for further analysis (Vickers, 2006). The usability of classification should be borne in mind when developing small area classifications. As discussed in Chapter 3, users may want to link their own ancillary data set to the area classification for further profiling and analysis. If the linked data spread well across the area classification clusters, they portend for further analysis. Table 9.7 shows the extent to which clusters are sized in terms of *barangays* and population.

As can be seen from Table 9.7, there is an unclassified cluster comprising 398 *barangays*. The reason for an unclassified cluster is that data were not provided for 14 *barangays*, while 384 *barangays* displayed characteristics that did not readily fit into any of the existing clusters.

Radial plots have been developed for all clusters at the three hierarchical levels of the classification. The radial plot of Enterprise Flux is shown in Figure 9.6. The plot shows continuous variables.

In addition to radial plots, pen portraits which are comprehensive textual descriptions of the different clusters have also been created for each cluster. Statistical jargon was minimized during the development of

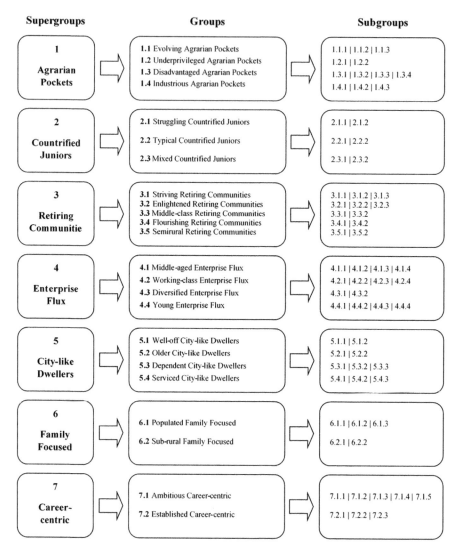

FIGURE 9.5
Hierarchical structure and labels of the Philippines clusters (Author's elaboration previously published in Ojo, A., Vickers, D. and Ballas, D. (2013). Creating a Small Scale Area Classification for Understanding the Economic, Social and Housing Characteristics of Small Geographical Areas in the Philippines. *Regional Science Policy and Practice*, 5(1), 1–24).

pen portraits. An extract from the pen portrait for Enterprise Flux is shown in Plate 9.1.

Due to the physical configuration of the Philippines, it is generally challenging to display static visualizations for the entire country at the *barangay* scale. Dynamic mapping platforms are generally better for visualizing

TABLE 9.7
Distribution of *barangays* and population in each cluster (2000)

Cluster	Barangays (count)	Barangays (%)	Population (count)	Population (%)
Agrarian Pockets	9,862	23.51	12,494,336	16.35
Countrified Juniors	5,233	12.48	8,802,716	11.52
Retiring Communities	6,919	16.50	10,365,522	13.57
Enterprise Flux	7,955	18.97	12,143,974	15.89
City-like Dwellers	5,751	13.71	20,648,202	27.02
Family Focused	2,258	5.38	4,036,362	5.28
Career-centric	3,564	8.50	6,953,227	9.10
Unclassified	398	0.95	969,318	1.27

Source: Author's calculation previously published in Ojo, A., Vickers, D. and Ballas, D. (2013). Creating a Small Scale Area Classification for Understanding the Economic, Social and Housing Characteristics of Small Geographical Areas in the Philippines. *Regional Science Policy and Practice*, 5(1), 1–24.

Plate 9.1 Extract of Pen Portrait for Enterprise Flux

Within the cluster, it is common to find people employed in various sectors of the economy. Professionals and senior officials are resident in high proportions. Sectors that employ people in large proportions include transport and communications, education, health and social care, real estate and business, manufacturing, and auto repair enterprises. Small and privately-owned establishments are also very common within these areas. People aged 21 and 44 years dominate these communities. Filipino speakers also tend to be resident at disproportionately high levels. The availability of elementary schools is above the national mean and a substantial number of residents have postsecondary technical education. Compared to the national mean, electricity supply is generally high and liquid fuels are often used for cooking. A substantial number of households have television sets and utilize water closet toilets. Houses are generally built with bricks, stones, or concrete. Mixed building materials are often used, and it is typical to find corrugated iron or concrete roofing within these areas. Southern Tagalog is the region with largest share of people resident within Enterprise Flux communities. It is followed by Central Luzon and Western Visayas, respectively. The province with the largest share of people resident within Enterprise Flux communities is the City of Marikina.

mapped data in the Philippines. Nevertheless, for the purpose of this chapter, an illustration of the spatial distribution of the area classification is shown for Central Luzon region in Figure 9.7.

9.7 Conclusion

Barangays are the smallest administrative unit of population aggregation in the Philippines. They therefore serve as an important basis for understanding

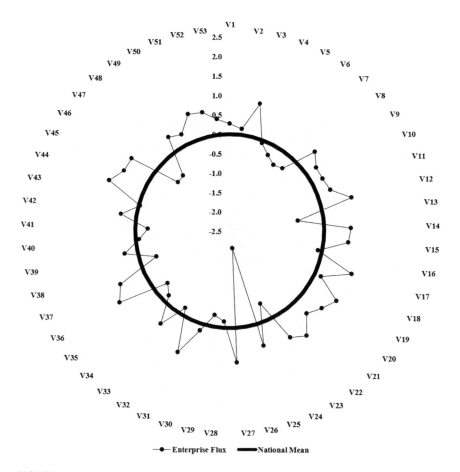

FIGURE 9.6

Radial plot of enterprise flux (Author's elaboration previously published in Ojo, A., Vickers, D. and Ballas, D. (2013). Creating a Small Scale Area Classification for Understanding the Economic, Social and Housing Characteristics of Small Geographical Areas in the Philippines. *Regional Science Policy and Practice*, 5(1), 1–24).

V1	Population density	V28	Banking institutions
V2	No married couples	V29	Recreation establishments
V3	Two or more married couples	V30	Disabled
V4	Stepchild of head	V31	Electricity supplied
V5	Age 0–5 years	V32	No piped water
V6	Age 6–10 years	V33	Cooking energy: liquid fuels
V7	Age 11–20 years	V34	Cooking energy: wood
V8	Age 21–44 years	V35	No telephone
V9	Age 45–64 years	V36	Television
V10	Age 65+ years	V37	Water closet toilet
V11	Widowed	V38	Latrine
V12	Speaks Filipino	V39	Dwelling owned
V13	Less than primary education	V40	Members squatting
V14	Postsecondary technical education	V41	Free occupancy
V15	Professionals and senior officials	V42	Land occupied with consent
V16	Agricultural and fishery workers	V43	Built with bamboo
V17	Transportation and communications employment	V44	Built with bricks, stone, or concrete
V18	Public administration and defense employment	V45	Built with mixed materials
V19	Education employment	V46	Iron and concrete roofing
V20	Health and social work employment	V47	Buddhist
V21	Overseas worker	V48	Muslim
V22	Real estate and business establishments	V49	Roman Catholics
V23	Private establishments	V50	Other Christians
V24	Cooperative establishments	V51	Manufacturing establishments
V25	Small establishments	V52	Auto repair shops
V26	Media establishments	V53	Restaurants and personal services
V27	Retail trade establishments		

FIGURE 9.6
(Cont.)

the local impact of national socioeconomic policies. However, it is very rare to find the type of extensive nationwide analysis conducted at the *barangay* scale. The analysis discussed in this chapter is therefore pathbreaking. One of the innovative aspects of the analysis is the combination of categorical and continuous data for developing a small area geodemographic classification. In addition to contributing to the scarce literature on open geodemographic

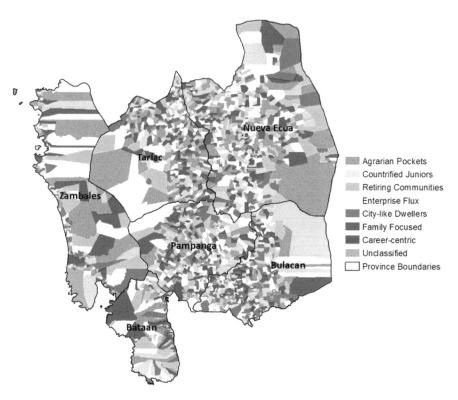

FIGURE 9.7
Map of central Luzon region showing the spatial distribution of supergroups.

classification in developing countries, the Philippines' area classification system will prove useful for addressing numerous policy-related issues.

References

Agresti, A. (2013). *Categorical Data Analysis*. Hoboken, NJ: John Wiley & Sons.

Agresti, A. (2018). *Statistical Methods for the Social Sciences*. London: Pearson.

Akaike, H., Parzen, E., Tanabe, K. and Kitagawa, G. (1998). *Selected Papers of Hirotugu Akaike*. New York, NY: Springer.

Banerjee, S., Carlin. B.P. and Gelfand, A.E. (2014). *Hierarchical Modeling and Analysis for Spatial Data*. Boca Raton, FL: CRC Press.

Coursey, O.W. (2008). *History and Geography of the Philippine Islands*. Charleston, SC: BiblioBazaar.

Crawshaw, J. and Chambers, J. (2001). *A Concise Course in Advanced Level Statistics with Worked Examples*. Cheltenham: Nelson Thornes.

Gordon, R.G., Jr. (Ed). (2005). *Ethnologue: Languages of the World*. Dallas, TX: SIL International.

Everitt, B.S., Landau, S. and Leese, M. (2001). *Cluster Analysis*. London: Arnold.

Hellerstein, J.M. and Stonebraker, M. (2005). *Readings in Database Systems*. Cambridge, MA: Massachusetts Institute of Technology Press.

Johnson, N.L., Kotz, S. and Balakrishnan, N. (1994). *Continuous Univariate Distributions*. New York, NY: John Willey & Sons.

Jolliffe, I.T. (2002). *Principal Components Analysis*. New York, NY: Springer.

Larose D.T. (2006). *Data Mining Methods and Models*. Hoboken, NJ: Wiley.

Maydeu-Olivares, A. and Joe, H. (2014). Assessing Approximate Fit in Categorical Data Analysis. *Multivariate Behavioral Research*, 49(4), 305–328.

NSO (2008). *The Philippines in Figures 2008*. Manila: National Statistics Office.

NSO (2010). *The Philippines in Figures 2010*. Manila: National Statistics Office.

Ojo, A., Vickers, D. and Ballas, D. (2013). Creating a Small Scale Area Classification for Understanding the Economic, Social and Housing Characteristics of Small Geographical Areas in the Philippines. *Regional Science Policy and Practice*, 5(1), 1–24.

Saxena, S. (2018). Drivers and Barriers to Re-Use Open Government Data (OGD): A Case Study of Open Data Initiative in Philippines. *Digital Policy, Regulation and Governance*, 20(4), 358–368.

UN (2019). *World Population Prospects 2019: Online Edition. Rev. 1*. New York, NY: United Nations Department of Economic and Social Affairs, Population Division.

Vickers, D.W. (2006). *Multi-Level Integrated Classifications Based on the 2001 Census*. PhD thesis, School of Geography, University of Leeds, United Kingdom.

Vickers, D. and Rees, P. (2006). Introducing the Area Classification of Output Areas. *Population Trends*, 125, 15–29.

Webber, R. (2004). *The Relative Power of Geodemographics vis a vis Person and Household Demographic Variables as Discriminators of Consumer Behavior*. London: Centre for Advanced Spatial Analysis.

WHO (2019). *World Malaria Report 2019*. Geneva: World Health Organization.

Woods, D.L. (2006). *The Philippines: A Global Studies Handbook*. Santa Barbara, CA: ABC-CLIO.

Worcester, D.C. (2005). *The Philippine Islands and Their People*. Boston, MA: Adamant Media Corporation.

10

Modeling Temporal Distribution and Seasonality of Infectious Diseases with Area Classifications

10.1 The Burden of Infectious Diseases

The aim of this chapter is to investigate the seasonal spatial patterns of an infectious disease in a tropical African country. Diseases may be described as disturbances that lead to disorders in the state of health of living organisms (Warrell et al., 2016). All living organisms are susceptible to experiencing some form of disturbance in their state of health. Diseases are caused when one organism (agent) invades a territory within another organism (host). Diseases are caused by parasitic and pathogenic organisms. Parasites are disease-causing organisms that live on or inside the body of other organisms and thrive to harm their host, while pathogens are defined as organisms causing diseases to their host (Vynnycky and White, 2010). As illustrated in Figure 10.1, there are numerous ways through which disease-causing agents prove detrimental to their hosts. Disease-causing agents steal space, nutrients, and living tissue from their symbiotic hosts.

From earliest records of human history, it is known that infectious diseases have shaped the human population development. Infectious diseases are disorders that are caused mainly by infectious agents such as viruses, bacteria, protozoa, fungi, helminths, ectoparasites, and arthropods (Vynnycky and White, 2010). Infectious diseases received urgent global attention toward the end of December 2019 following the disclosure that a cluster of viral phenomena traced to a wet market in Wuhan (Hubei, China) had broken out (Wang et al., 2020). The coronavirus disease 2019 (COVID-19) outbreak was declared a global pandemic by the World Health Organization (WHO) in March 2020. The COVID-19 pandemic is a stark reminder that infectious diseases ignore national geographic and political boundaries. Consequently, these diseases should be treated as global threats that place all people everywhere at risk. Their rapid and uncontrolled outbreaks can quickly overwhelm

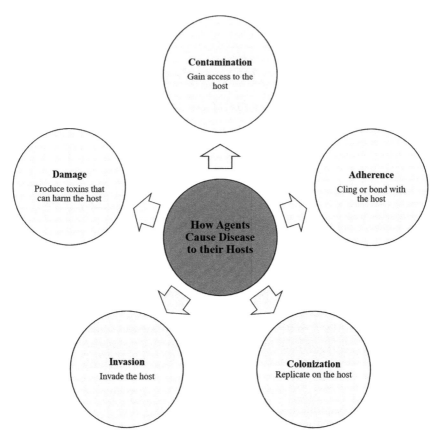

FIGURE 10.1
How parasites and pathogens cause disease.

health systems, resulting in limited capability to handle routine healthcare demands and an escalation in healthcare costs.

In addition to direct public health consequences, infectious diseases also generate social and economic shocks especially for the most vulnerable members of society. When infectious diseases overwhelm the societies, they force patients and those who take care of them in informal healthcare settings like homes to stay away from work. This generates additional productivity strain for the labor market. Some caretakers also end up losing the jobs in extreme situations. In epidemic or pandemic situations, homeless people can also be badly affected. Due to lack of access to adequate shelter, homeless people could be disproportionately exposed to the dangers of contracting infectious disease. Similarly, refugees and displaced persons can suffer heavy casualties from infectious disease outbreaks due to limited capability to move and restricted access to healthcare facilities. Perry and Donini-Lenhoff (2010) documented detailed evidence showing that social stigmatization has been and is still a major problem that complicates the management of infectious diseases. Continuous education is the key to combating both the spread of infectious diseases and the stigmatization of sufferers of these diseases.

Although infectious diseases are found in every region of the world, in relative and absolute terms, they represent a considerably higher burden in developing countries than in developed countries (Bygbjerg, 2012). To get a sense of the scale of infectious diseases burden within more developed and less developed parts of the world, Figure 10.2 presents data extracted from the Global Burden of Disease Collaborative Study 2017 (GBD 2017). The GBD study is coordinated by the Institute for Health Metrics and Evaluation (IHME), which is based in Seattle, USA. Figure 10.2 shows death rates for infectious, maternal, neonatal, and nutritional diseases. The chart compares global rates with groups of countries based on their sociodemographic index (SDI). According to the IHME, SDI is a summary measure that identifies where countries sit on the spectrum of development. SDI is a composite average of the rankings of the per capita income, average educational attainment, and fertility rates of all areas in the GBD study.

Figure 10.2 clearly signifies that the rate of these diseases is in the downward direction, although evid ence from the same data set used to produce the chart shows that the number of cases has been rising. Low and low-middle

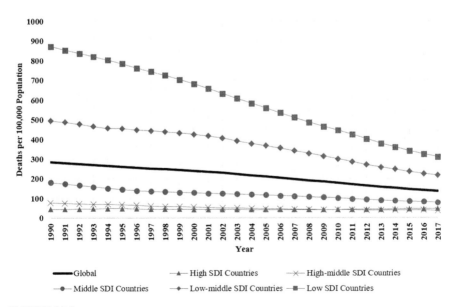

FIGURE 10.2
Death rates for infectious, maternal, neonatal, and nutritional diseases (Author's elaboration based on data from the Global Burden of Disease Study 2017, Institute for Health Metrics and Evaluation (IHME)).

SDI countries feature a vast number of developing countries, and as can be seen from the chart, rates of the diseases in these areas of the world are much higher than that of other regions and the entire world. Infectious diseases are a leading cause of death worldwide, particularly in low-income countries, especially in young children. At the beginning of the current millennium, infectious diseases accounted for six in ten childhood deaths and four in ten premature deaths (WHO, 2000).

10.2 Importance of Surveillance

Rapid detection and reporting of infectious diseases are very essential in curtailing the spread of these diseases among susceptible population groups. Effective surveillance systems rely on good laboratory testing facilities, human resources, and processes, many of which are in short supply in developing countries (Jamison et al., 2006). Consequently, a vast number of deaths are triggered by epidemic infectious diseases such as cholera, meningococcal disease, and measles. This chapter uses a geodemographic area classification to explore the spatial burden of measles in a developing country such as Nigeria.

Measles is a highly contagious infectious and vaccine-preventable disease that causes periodic seasonal epidemics and shows marked seasonality in some countries (Conlan and Grenfell, 2007). The disease is caused by *Morbillivirus* of the Paramyxoviridae family (Diallo, 1990) and it is a contagious respiratory infection. Prior to the introduction of a vaccine in 1963 (WHO, 2012), major measles epidemics happened roughly every two to three years, resulting in an estimated 2.6 million deaths annually. In 2018, more than 140,000 people died from measles (WHO, 2019). The WHO has recommended molecular surveillance of measles strains and established a vaccination program for measles elimination (WHO, 2012).

At the global level, the prime objective of surveillance is to identify areas of measles virus transmission and immunity gaps in order to guide effective public health responses. There are four broad objectives of surveillance at national and local levels. These include:

- Detection and confirmation of cases to ensure proper case management and implementation of appropriate public health strategies
- Investigation of cases to determine the source of infection
- Identification of populations and areas with low coverage and at higher risk of outbreaks that require enhanced vaccination efforts
- Verification of the absence of endemic measles cases to document elimination of endemic virus.

The global and local objectives of the WHO surveillance strategy underscore the importance of conducting local level spatial analysis.

10.3 Data Modeling Approach

The spread of measles virus in communities has been found to be influenced not only by the veracity of the virus, but also by factors such as age, immunological conditions of persons, physical factors, indoor crowding, school activity, and social contacts (Kouadio, 2010; Nsubuga et al., 2018). Selected recent descriptive studies on the epidemiology of measles in Nigeria have been published for parts of country (Fatiregun and Odega, 2013; Isere and Fatiregun, 2014; Shorunke et al., 2019). However, until now, no major studies have focused on the relationship between the burden of measles and population-level characteristics of communities during different seasons. Studying population-level and seasonal variation in the burden of measles is a potentially useful way to strengthen surveillance because it recognizes contextual-level influences. By combining measles mapping and the Nigerian geodemographic area classification, the study reported in this chapter aims to profile communities where measles frequently or infrequently occurs during specific seasons.

10.3.1 Source of Infectious Disease Data Set

Data on infectious diseases from 2006 to 2018 were obtained from the National Disease Outbreak Dashboard of the Nigeria Centre for Disease Control (NCDC). Although the NCDC was conceived in 2007 as an attempt to establish an institution that can effectively mobilize resources to respond to disease outbreaks, it was formally established in 2011. The data set comprised 40,088 records with unique subject identifiers of cases of persons who contracted six types of infectious diseases, namely, cholera, cerebrospinal meningitis, Lassa fever, measles, monkeypox, and yellow fever.

10.3.2 Nigerian Seasons

Nigeria is wholly situated within the tropics and there are two distinct seasons: a wet and a dry season. The wet season spans from April to October, with generally lower temperature (Falola et al., 2019). The dry season spans from November to March, with midday temperatures that surpass 38°C (100°F) combined with relatively cool nights (Falola et al., 2019). Both seasons are characterized by long and short durations, resulting in four seasonal intervals described below..

Long Rainy Season: This starts in March and lasts till the end of July, with a peak period in June over most parts of southern Nigeria. It is a period of thick clouds and is excessively wet, particularly in the Niger Delta and the coastal lowlands. It is marked by humidity with values hardly below 85% in most parts of the forested south.

Short Dry Season: This is experienced in August for three to four weeks. However, the main dry period known as the "August break" is generally observed in the final two weeks of August in most parts of southern Nigeria.

Short Rainy Season: This brief wet period follows the "August break" from early September to mid-October, with a peak period at the end of September. The rains are not usually as heavy as those in the long rainy season, although the spatial coverage over southern Nigeria is similar. The two periods of rainfall intensity give the double maxima phenomenon of the rainy season characteristic of southern Nigeria. The short dry season in August between these two rainy periods allows for harvesting and planting of fast-growing varieties of grains, such as maize.

Long Dry Season: This period starts from late October and lasts till early March, with peak dry conditions between early December and late February. The period witnesses the prevailing influences of the dry and dusty northeast winds, as well as the "harmattan" conditions. Vegetation growth is generally hampered, grasses dry, and leaves fall from deciduous trees due to reduced moisture.

10.3.3 Modeling Approach

One of the long-standing challenges confronting Nigerian public health officials is the understanding of local level seasonal variations in infectious disease patterns as well as their population level correlates for effective surveillance and deployment of resources and public health interventions. By integrating GIS with geodemographic area classifications, it is possible to compile and track information about the burden and spread of a disease (Abbas et al. 2009). Furthermore, as Forcen and Salazar (2000) rightly observed, the characteristics of a place, including its population characteristics and environment, are often critical factors in determining the origin and spread of a disease and may offer insights into its prevention and control.

Each case in the infectious disease data file contained a date of onset, which signifies the date of first appearance of the signs or symptoms of the illness. The date of onset was used to allocate each case to one of the four seasons described in the previous section. The data file also contained corresponding residential LGAs of included cases. A lookup file was used to link the LGAs in the infectious disease data file with the Nigerian geodemographic classification previously described in Chapter 8 of this book.

The epidemiologic condition of interest in this chapter is morbidity, which is essentially the state of being diseased. The first objective was to understand the temporal distribution of measles cases relative to temperature and precipitation. Community-based distribution of the disease was juxtaposed against monthly average temperature and precipitation data recorded between 2006 and 2016. Monthly temperature and precipitation values were derived from the World Bank Climate Change Knowledge Portal (CCKP). The portal provides an online platform for access to comprehensive global, regional, and country data related to climate change and development.

The second objective was to focus on the seasonality of the burden of measles and seek to answer the question, "what is the spatial configuration and contextual descriptors of communities that are likely to experience a higher prevalence of measles during specific seasons?" Answers to this type of question can be particularly useful for measles surveillance and response strategy.

Prevalence is the proportion of a population who have a specific characteristic in a given time period (Williams, 2001). There are several ways to estimate and report prevalence depending on the timeframe of the estimate. Point prevalence is the proportion of a population that has the disease at a specific point in time. Period prevalence is the proportion of a population that has the disease at any point during a given time period of interest. Lifetime prevalence is the proportion of a population who, at some point in life, have ever had the disease.

$$MI_k = \frac{\frac{n}{\sum_1^k n}}{\frac{N}{\sum_1^k N}} \tag{10.1}$$

where:

MI_k represents the relative prevalence of measles for geodemographic group k

n represents the number of measles cases recorded for geodemographic group k

N represents the population at risk in geodemographic group k.

Period prevalence provides better measure of the disease load because it incorporates all new cases and all deaths between two dates, whereas point prevalence only counts those alive on a particular date. An index value shown in Equation 10.1 was calculated and used to summarize the prevalence of measles within each geodemographic group, relative to the prevalence of the disease across all groups and standardized against a score of 100. When the value of MI_k in a geodemographic group is 100, the relative prevalence of incidence of measles in this group is the same as that expected, and a value of 200 indicates that the observed number of cases is 100% higher than that expected in the study area.

10.4 Community-Based Distribution of Measles and Meteorological Factors

Recorded measles cases accounted for approximately 30% of all infectious disease cases within the NCDC data set. Since the temperature and precipitation data secured from the CCKP were recorded between 2006 and 2016, measles cases during the same time period were used for this part of the analysis. Although the total number of clinically recorded cases in the NCDC dashboard during 2006 to 2016 was 10,972, this figure grossly underestimates the actual number of measles cases in Nigeria. Overall, seven in ten cases of measles were reported for children under five years old. Measles occurrence remains a leading cause of mortality especially among children under five years of age (Dabbagh et al., 2018).

Table 10.1 shows the average monthly temperature and precipitation statistics recorded for each month during the 2006 to 2016 period, along with the cumulative percentage shares of recorded measles cases during the same period. Around 55% of cases occurred during the first six months and 45% of cases reported in the latter half of the year. The largest share of measles cases during the period (14%) occurred in the month of March, which also reported one of the highest average temperatures.

The results presented in Table 10.1 only tell a part of the story. It is important to interpret the distribution of cases within the sociodemographic context of communities. To get a little more clarity on the relationship between the monthly distribution of measles, the two physical factors, and the social

TABLE 10.1
Rainfall, temperature, and cumulative share of measles cases in Nigeria (2006–2016)

Month	Average Rainfall (MM) (2006–2016)	Average Temperature (Celsius) (2006–2016)	Measles Cases (%) (2006–2016)
January	2.82	25.24	6.30
February	8.03	28.15	10.84
March	25.96	30.20	14.29
April	53.75	30.60	8.75
May	112.78	29.54	7.79
June	161.35	28.04	6.54
July	182.14	26.58	4.82
August	232.27	25.68	8.08
September	218.24	26.32	6.25
October	109.53	27.46	10.28
November	12.85	26.89	9.34
December	4.15	25.25	6.72

Source: Author's calculation based on data from the World Bank CCKP and the NCDC dashboard.

context of communities, the share of cases is calculated by geodemographic typology. Results summarized in Figure 10.3 are shown for normalized values. All values were normalized using the range normalization method discussed in Chapter 5. Figure 10.3 confirms that the temporal distribution of measles cases within different types of communities is associated with temperature and precipitation in different ways. The results point to nonlinear relationships between the two meteorological variables and the distribution of measles by community types.

Basically, measles cases rise and fall with temperature within Emerging Localities and Country Dwellings. However, the trend of measles cases within Emerging Localities most closely followed the amplitude of average temperature during the study period, yielding a correlation coefficient of 0.5. Emerging Localities are largely concentrated in the northwest and northeast pockets of Nigeria. In comparison with the other five community types, Emerging Localities are also characterized by a disproportionate concentration of young people. The cumulative share of measles cases within Emerging Localities in September however bucks this trend of cases rising and falling with temperature. We see a sharp rise in the distribution of cases in September despite the dip in temperature. This is probably associated with the academic

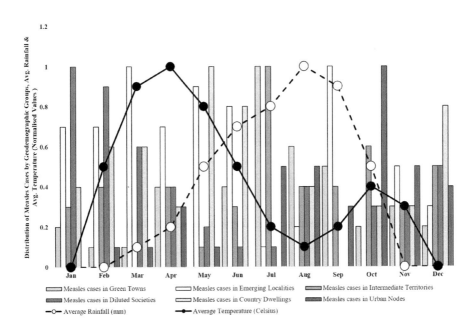

FIGURE 10.3
Relationship between geodemographic distribution of measles cases and physical factors (2006–2016) (Author's elaboration based on data from the Nigerian geodemographic classification, the World Bank CCKP, and the NCDC dashboard).

calendar of primary schools. Public primary schools in Nigeria generally resume in September, which may stimulate the rise in numbers of infections.

When one observes the amplitude of average monthly rainfall, and the distribution of measles cases within geodemographic typologies, a totally different pattern emerges. Measles cases rise and fall with rainfall within Green Towns, Intermediate Territories, and Urban Nodes.

The strongest correlation between measles distribution and precipitation is noticeable within Green Towns (Spearman's correlation coefficient of 0.7). Green Towns concentrate mostly in the southwestern corner of Nigeria and can also be found in the north central, south-south, southeast and northeast geopolitical zones. On average, population density within these communities is about 450 people per square kilometer and they also tend to be comprised of residents in older age categories with pensioners.

With a mean household size of 4.6 persons and a very high population density of 5,117 persons per square kilometer, Urban Nodes are scattered across the country and do not necessarily concentrate in any geopolitical zone. However, the North East has the lowest share of Urban Nodes. Generally, measles cases in Urban Nodes rise and fall with rainfall as can be observed in Figure 10.3. However, there is an anomaly in the month of October as the dip in precipitation does not translate into a corresponding fall in measles cases. As a matter of fact, the cumulative number of measles cases within Urban Nodes during the study period peaked in the month of October. The reasons for this increased morbidity are not immediately apparent from the analysis. The spike may be linked to the commemoration of Nigeria's independence from Britain, which mainly takes place in cities in October. During the celebration, people are out in the streets and several schools also put on their own parades, thereby increasing contact among school-aged children. An alternative hypothesis is that the October spike in Urban Nodes may be connected with the presence of urban slums and rural–urban migratory patterns in majority of cities. The slums in Urban Nodes facilitate overcrowding that has been identified as a crucial risk factor in the transmission of respiratory infections, including measles (Baker et al., 2001). If the migration into urban slums increases during the October period, there is the potential for the number of cases to increase within Urban Nodes. However, this line of reasoning requires further investigation.

10.5 Spatial Targeting of Resources Based on Seasonal Disease Load

The discussion presented in Section 10.4 provides an understanding of the community-level variations in the frequency of measles that exist in Nigeria. Certain geodemographic factors seem to be associated with these variations.

In addition, we also discover from the discussion that there may be a link to how different community types respond to the meteorological conditions associated with the distribution of measles. Therefore, for the purpose of spatially targeting interventions, it makes sense to investigate the load of the disease on a seasonal basis.

The discussion reported in this section represents the first opportunity to explore the socioeconomic profile of communities in Nigeria and the extent to which this profile explains variations in the spatial configuration of the load of measles during different seasons. The Nigerian geodemographic classification of 774 LGAs, which was discussed in Chapter 8, revealed 6 clusters at the first level (supergroups) and 23 clusters at the second level (groups) of the hierarchy. The 23 groups are used here to illustrate variations in the seasonal prevalence of measles within each geodemographic group (2006 to 2018), relative to the prevalence of the disease across all groups and standardized against a score of 100. The analysis of differential measles outcomes across these 23 geodemographic clusters revealed clear geographical health inequalities.

Table 10.2 shows an example of disease profiles by geodemographic clusters. The table compares the relative period prevalence of measles across all seasons with the prevalence of the disease for each of the four key seasons previously described in this chapter. With regard to the load of measles across all seasons, Table 10.2 shows that although Comfortable Emerging Localities account for just 5.6% of the population at risk, one can expect over two times the average prevalence of measles in these areas. Comfortable Emerging Localities make up 45 LGAs in 10 states (Zamfara, Gombe, Jigawa, Kaduna, Kano, Katsina, Kebbi, Niger, Sokoto, and Bauchi). Access to health services is below the national mean, with a large number of households spending over an hour on travel to reach their nearest health facility. Those who have access to health services are often dissatisfied with the absence of trained professionals and lack of medicines. As a result, many people dwell on the use of orthodox and traditional medicine. Some people also patronize local chemists or pharmacists.

LGAs classified as Deprived Diluted Societies however experience less than half of the average for the entire country. Deprived Diluted Societies can be found in 5 states (Bauchi, Benue, Taraba, Plateau, and Ebonyi) and they spread across 13 LGAs. The levels of economic activity are very low within these communities. There are below average representations of self-employed people and residents are predominantly employed in the agricultural sector. Unemployment rate is just above the national average.

From column 4 in Table 10.2, one can discover that the ratio of the highest disease-loaded communities to the lowest disease-loaded communities is approximately 12:1 across all seasons. This has obvious implications for public health strategies and resource targeting.

Figure 10.4 highlights those areas where the population at risk have the potential to experience high, medium, and low loads of measles irrespective

TABLE 10.2

Geodemographic profile of seasonal load of measles in Nigeria (2006–2018)

Geodemographic Group	Population at Risk (%)	All Seasons		Long Rainy Season		Short Dry Season		Short Rainy Season		Long Dry Season	
		Measles Cases (%)	Relative Prevalence Index	Measles Cases (%)	Relative Prevalence Index	Measles Cases (%)	Relative Prevalence Index	Measles Cases (%)	Relative Prevalence Index	Measles Cases (%)	Relative Prevalence Index
Conventional Green Towns	4.86	2.66	55	2.58	53	1.99	41	2.83	58	2.88	59
Underprivileged Green Towns	3.48	3.94	113	3.84	110	4.46	128	4.67	134	3.59	103
Flourishing Green Towns	5.08	3.69	73	3.43	68	4.84	95	3.53	70	3.79	75
Struggling Green Towns	6.92	4.51	65	4.51	65	6.55	95	4.95	71	3.74	54
Moderately Emerging Localities	8.36	17.89	214	19.14	229	27.64	331	16.63	199	14.11	169
Comfortable Emerging Localities	5.59	13.82	247	19.45	348	5.41	97	6.58	118	11.99	214
Transient Emerging Localities	7.26	8.18	113	5.78	80	6.46	89	9.73	134	11.15	154
Constrained Intermediate Territories	1.30	0.92	70	0.58	44	1.14	87	1.74	133	0.92	70
Well-to-do Intermediate Territories	6.64	4.46	67	4.24	64	3.51	53	5.38	81	4.58	69
Deprived Intermediate Territories	1.75	0.98	56	0.85	49	0.57	33	0.92	53	1.31	75
Customary Intermediate Territories	3.08	3.01	98	2.87	93	3.42	111	3.10	101	3.04	99
Thriving Diluted Societies	3.27	2.76	85	2.16	66	3.51	107	1.79	55	3.85	118
Laboring Diluted Societies	5.35	2.03	38	1.29	24	0.85	16	0.82	15	3.93	73
Deprived Diluted Societies	1.50	0.30	20	0.39	26	0.09	6	0.16	11	0.31	21
Modest Diluted Societies	4.14	1.54	37	1.99	48	0.85	21	1.25	30	1.26	30
Toiling Country Dwellings	2.38	4.25	179	5.80	244	0.66	28	1.96	82	4.24	178
Deprived Country Dwellings	0.74	0.39	53	0.27	36	0.66	90	0.43	59	0.47	64
Middle-Class Country Dwellings	6.48	8.75	135	8.31	128	9.31	144	7.23	111	9.92	153
Prosperous Urban Nodes	2.58	1.04	40	0.96	37	1.61	63	1.20	46	0.92	36
Disadvantaged Urban Nodes	2.97	2.84	96	2.26	76	3.13	106	2.99	101	3.48	117
Average Urban Nodes	3.46	5.79	167	4.22	122	6.08	176	13.21	381	4.27	123
Affluent Urban Nodes	10.39	4.82	46	3.82	37	6.17	59	7.39	71	4.58	44
Striving Urban Nodes	2.42	1.42	59	1.25	52	1.04	43	1.52	63	1.70	70

Source: Author's calculation based on data from the Nigerian geodemographic classification and the NCDC dashboard.

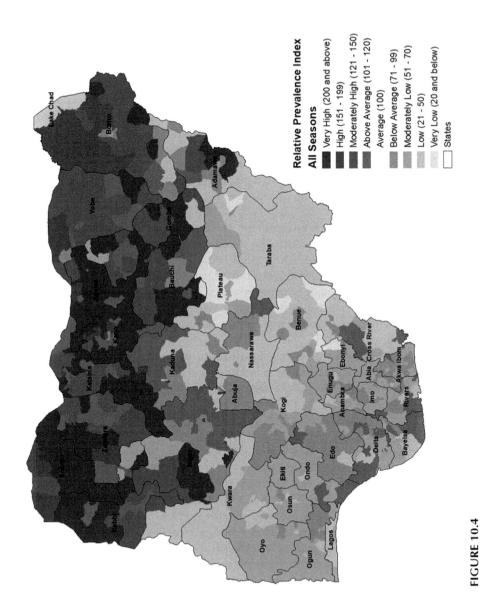

FIGURE 10.4

Load of measles during all seasons (2006–2018).

of the season. These spatial divisions are an extrapolation of the prevalence analysis reported in Table 10.2. Similar maps are shown in Figures 10.5 to 10.8 for each of the different seasons analyzed. These maps demonstrate the usefulness of this approach for spatially targeting public health interventions especially in resource-poor settings. By highlighting those local areas with a high load of measles during different seasons, the results could assist both policy and sensitization efforts in these areas.

Columns five and six in Table 10.2 identify those occurrences of measles during the long rainy season which starts in March and continues to the end of July. Here, Laboring Diluted Societies, Deprived Diluted Societies, Prosperous Urban Nodes, and Affluent Urban Nodes all have disease loads of a similar magnitude, which is below the national average. Many of the community types have values beneath the national average, but once again Comfortable Emerging Localities appear to experience a load well above the average load (three times that of the national average), indicating a higher propensity for the population at risk to experience measles during the long rainy season.

The seventh and eighth columns in Table 10.2 record measles community variations during the short dry season that takes place in August for three to four weeks. Here, Moderately Emerging Localities are particularly predisposed to measles during this season. Wells (particularly unprotected wells) are the major sources of drinking water within these areas. Furthermore, the mean household size is roughly 5.4 persons, while the population density is significantly below the national average, at 234 persons per square kilometer.

With regard to the load of measles during the short rainy season, it can clearly be seen that Average Urban Nodes experience a much higher prevalence. Population at risk within these communities are nearly four times likely to be loaded by measles during the brief wet period that follows the "August break" (early September to mid-October). Average Urban Nodes can be found in 6 states (Sokoto, Kaduna, Kano, Katsina, Niger, and Gombe) and the spread across 16 LGAs. Among the five Urban Nodes, Average Urban Nodes have the largest representation of young children. Although the overall level of access to health is slightly higher than the national average, the absence of trained professionals and long waiting times are major factors that make residents dissatisfied with health service. Private and religious hospitals are commonly used within Average Urban Nodes. From column 10 in Table 10.2, one can observe that the ratio of the highest disease-loaded communities to the lowest disease-loaded communities is approximately 35:1 during the short rainy season. Again, this represents a massive disparity across community types.

The long dry season takes place from October till early March and is characterized by the prevailing influences of dry and dusty northeast wind. During this season, there is a very high propensity for areas labeled as Comfortable Emerging Localities to display a very high prevalence of measles

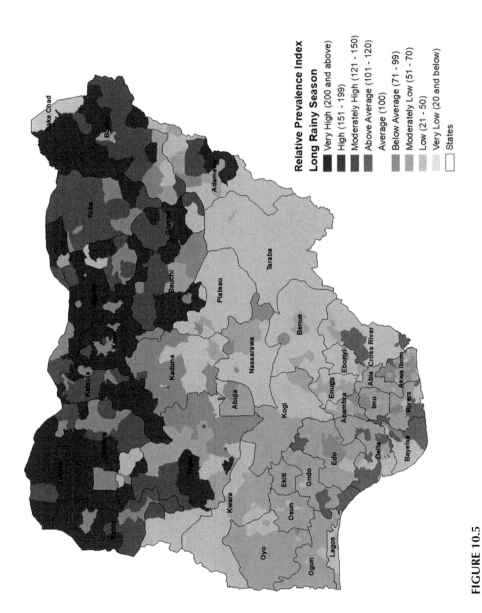

FIGURE 10.5

Load of measles during the long rainy season (2006–2018).

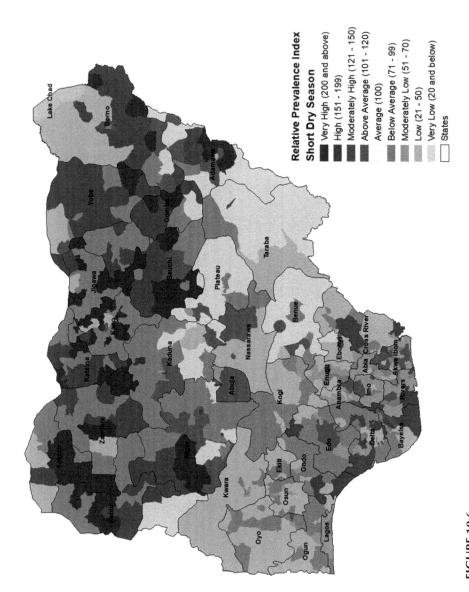

FIGURE 10.6
Load of measles during the short dry season (2006–2018).

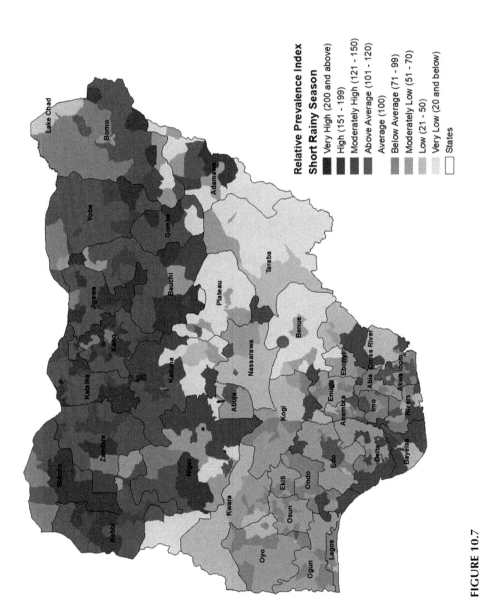

FIGURE 10.7

Load of measles during the short rainy season (2006–2018).

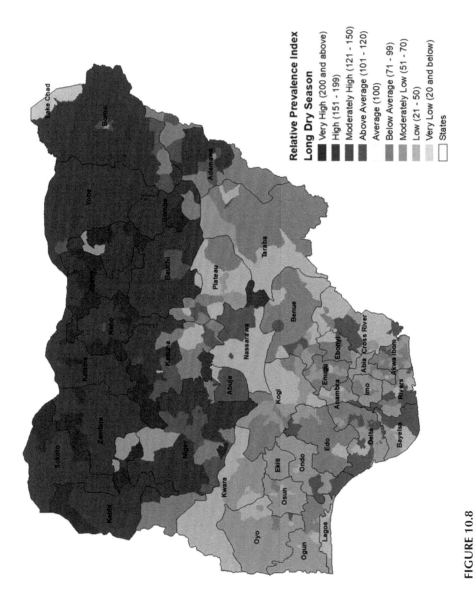

FIGURE 10.8

Load of measles during the long dry season (2006–2018).

(double that of the national average). Conversely, the prevalence index (i.e., less than half of the national average) recorded within Modest Diluted Societies during this season is the lowest across all geodemographic clusters.

10.6 Conclusion

In addition to adequate community awareness and strong social mobilization, surveillance to assess the burden of disease and guide interventions remains critical for the eradication of measles in Nigeria. Given scarce resources, the seasonal targeting of interventions needs to be conducted systematically to improve population health outcomes. The spatial patterns illustrated in this chapter may be used alongside the geodemographic profiles, by local, state, and federal public health agencies for surveillance, programming, and planning. The arguments are generally hermeneutic since the research findings should be construed in light of additional knowledge in the public health agencies, which is beyond the scope of this chapter. The issues confronting the health sector in most developing countries are complex and varied. Research-oriented solutions will therefore require an integrative multidisciplinary approach underpinned by collaborative partnerships among healthcare specialists and allies from other disciplines. The discussion summarized in this chapter further shows that the deployment of spatial analytical techniques within a GIS and geodemographics framework in the developing world holds monumental promise for answering complex ecological questions in health promotion, public health, community medicine, and epidemiology. Geodemographic frameworks should also offer health program managers the opportunity to better conduct program evaluation, develop operational health metrics, predict outbreaks of infectious diseases, and conduct health market demand and supply analysis.

References

Abbas, J., Ojo, A. and Orange, S. (2009). Geodemographics: A Tool for Health Intelligence. *Public Health*, 123(1), 35–39.

Baker, M., Goodyear, R. and Howden-Chapman, P. (2001). *Household Crowding and Health*. Wellington: Housing Corporation.

Bygbjerg, I.C. (2012). Double Burden of Noncommunicable and Infectious Diseases in Developing Countries. *Science*, 337(6101), 1499–1501.

Conlan, A.J.K. and Grenfell, B.T. (2007). Seasonality and the Persistence and Invasion of Measles. *Proceedings of the Royal Society B*, 274(1614), 1133–1141.

Dabbagh, A., Laws, R.L., Steulet, C., Dumolard, L., Mulders, M.N., Kretsinger, K., Alexander, J.P., Rota, P.A. and Goodson, J.L. (2018). Progress Toward Regional Measles Elimination – Worldwide, 2000–2017. *Morbidity and Mortality Weekly Report*, 67(47), 1323–1329.

Diallo, A. (1990). Morbillivirus Group: Genome Organisation and Proteins. *Veterinary Microbiology*, 23(1–4), 155–163.

Falola, T.O., Kirk-Greene, A.H.M., Ajayi, J.F.A. and Udo, R.K. (2019). Nigeria. *Encyclopedia Britannica*, Chicago, IL: Encyclopedia Britannica, Inc.

Fatiregun, A.A. and Odega, C.C. (2013). Representativeness of Suspected Measles Cases Reported in a Southern District of Nigeria. *Asian Pacific Journal of Tropical Medicine*, 6(2), 131–134.

Forcen, E. and Salazar, A. (2000). GIS: An Effective Tool for Disease Monitoring. *GEOEurope*, 9(2), 31–32.

Isere, E.E. and Fatiregun, A.A. (2014). Measles Case-Based Surveillance and Outbreak Response in Nigeria: An Update for Clinicians and Public Health Professionals. *Annals of Ibadan Postgraduate Medicine*, 12(1), 15–21.

Jamison, D.T., Breman, J.G., Measham, A.R., Alleyne, G., Claeson, M., Evans, D.B., Jha, P., Mills, A. and Musgrove, P. (Eds). (2006). *Disease Control Priorities in Developing Countries*. Washington, DC: The World Bank.

Kouadio, I.K., Kamigaki, T. and Oshitani, H. (2010). Measles Outbreaks in Displaced Populations: A Review of Transmission, Morbidity and Mortality Associated Factors. *BMC International Health Human Rights*, 10(5). DOI: https://doi.org/10.1186/1472-698X-10-5.

Nsubuga, F., Bulage, L., Ampeire, I., Matovu, J.K.B., Kasasa, S., Tanifum, P., Riolexus, A.A. and Zhu, B. (2018). Factors Contributing to Measles Transmission During an Outbreak in Kamwenge District, Western Uganda, April to August 2015. *BMC Infectious Diseases*, 18(21). DOI: https://doi.org/10.1186/s12879-017-2941-4.

Perry, P. and Donini-Lenhoff, F. (2010). Stigmatization Complicates Infectious Disease Management. *Virtual Mentor*, 12(3), 225–230.

Shorunke, F.O., Adeola-Musa, O., Usman, A., Usman, A., Ameh, C., Waziri, E. and Adebowale, S.A. (2019). Descriptive Epidemiology of Measles Surveillance Data, Osun State, Nigeria, 2016–2018. *BMC Public Health*, 19, 1636.

Vynnycky, E. and White, R.G. (2010). *An Introduction to Infectious Disease Modelling*. Oxford: Oxford University Press.

Warrell, D., Cox, T., Firth, J. and Dwight, J. (Eds). (2016). *Oxford Textbook of Medicine: Cardiovascular Disorders*. Oxford: Oxford University Press.

Wang, C., Horby, P.W., Hayden, F.G. and Gao, G.F. (2020). A Novel Coronavirus Outbreak of Global Health Concern. *The Lancet*, 395(10223), 470–473.

WHO (2000). *WHO Report on Global Surveillance of Epidemic-Prone Infectious Diseases*. Geneva: World Health Organization.

WHO (2012). *Global Measles and Rubella Strategic Plan: 2012–2020*. Geneva: World Health Organization.

WHO (2019). *More than 140,000 Die from Measles as Cases Surge Worldwide* [Press release]. 5 December. Available at: www.who.int/news-room/detail/05-12-2019-more-than-140-000-die-from-measles-as-cases-surge-worldwide. Accessed on 23/03/2020.

Williams, H. (2001). Disease Definition and Measures of Disease Frequency. *Journal of the American Academy of Dermatology*, 45(1), S33–S36.

11

Segmenting Gender Gaps in Levels of Educational Attainment

11.1 Educational Attainment and Gender Parity

Education is recognized as an important tool for transforming lives, building peace, eradicating poverty, and facilitating sustainable development (Tilbury et al., 2002). The fourth SDG focuses on ensuring inclusive and equitable quality education and promoting lifelong learning opportunities for all. The adoption of this goal is a demonstration of political will to support transformative actions for inclusive, equitable, and quality education for all.

The large concentration of young people in developing countries can translate into a demographic dividend by providing them with quality education and necessary skills. At least three dimensions of education are proven to be integral for progress in developing countries. These include:

- Access to education
- Progression and performance outcomes
- Attainment.

Lewin (2015, p. 29) suggested that access to education includes "on-schedule enrolment and progression at an appropriate age, regular attendance, learning consistent with national achievement norms, a learning environment that is safe enough to allow learning to take place, and opportunities to learn that are equitably distributed". Despite global progress in access to basic education, developing countries lag behind the rest of the world (Little and Lewin, 2011). Many studies have explored the barriers working against school attendance in developing countries (King and Hill, 1993; Mertaugh et al., 2009). Such barriers include family influence, social class, parental attitudes, poverty, armed conflict, lack of access to transportation, and child labor.

In addition to understanding the barriers working against school attendance, studies have also explored the drivers of progression and performance outcomes (Damon et al., 2016). Knowledge of these can help education policymakers redefine their strategies toward improving equity in outcomes. Studies have shown that children from low-income families typically live in homes and attend schools with few educational resources (Halle, 1997; Chiu, 2007). Consequently, interventions have typically focused on increasing resources at school and at students' homes. Additional measures include the expansion of instructional time and addressing individual learning needs.

Educational attainment is defined as the highest grade completed within the most advanced level attended in the educational system of the country where the education was received (UN, 1998). While access to education together with progression and performance outcomes in developing countries has received greater attention from researchers, a few studies have examined the dynamics of educational attainment, especially the gendered dimensions.

The human capital base of a country is integral to its social and economic progress. One way to monitor the strength of a country's human capital is by using the educational attainment statistic. In addition, indices of educational attainment feed into other measures such as literacy, unemployment, household income, poverty, and crime. Some scholars have suggested that nations that perform well in educational attainment often have better social polarization outcomes (Dorling and Woodward, 1996; Dyer, 2010).

In much of the educational literature, the popular argument with relation to gender gaps in attainment and schooling in developing countries suggests widespread discrimination against females (Grant and Behrman, 2010). However, it is not clear whether this hypothesis holds true in developing countries with high literacy rates. The Philippines is a high-literacy developing country. The adult literacy rate of the country increased from 93.6% in 1990 to 98.2% in 2015, growing at an average annual rate of 0.81% (UIS, 2020). Dating back to the early 1980s, evidence-based research has shown that the education of women in the Philippines contributes immensely to the country's economic growth (Crawford and Sidener, 1982). The educational attainment of women is also likely to impact significantly the health and livelihoods of their children (Caldwell, 1981). Becker et al. (1993) also found that the educational attainment of women in the Philippines greatly influenced their health choices and behaviors, particularly with regard to maternal and child health. Their research also provided evidence of an urban–rural dichotomy in varying impacts of female educational attainment.

In a bid to improve the access of women to education and mitigate the challenges of gender parity within the educational sector, the government of the Philippines in 1987 introduced an affirmation of the equality of all citizens irrespective of their gender within the national constitution. The remainder of this chapter uses the National Capital Region (NCR) of the Philippines to illustrate how geodemographics might be used to contrast spatial aspects of gender gaps in educational attainment at primary, secondary, and higher levels.

11.2 Data Sets

Educational attainment data for 2003 were derived from the 2003 Demographic and Health Survey (DHS). The principal objective of the DHS is to "to provide up-to-date information on population, family planning, and health to assist policymakers and program managers in evaluating and designing strategies for improving health and family planning services in the country" (NSO and ORC Macro, 2004, p. 2).

The 2003 DHS is a nationally represented survey of 12,586 households. Although more than 60,000 individuals (men, women, and children) were involved as participants, the analysis presented in this chapter draws upon data for women aged 15 to 49 years and men aged 15 to 54 years.

The 2003 DHS, which is the eighth in a series of such surveys, is a collaborative effort between the Philippines NSO and the US government. It was conducted between June and September 2003, with financial support from the United States Agency for International Development (USAID) and technical inputs from ORC Macro, a research and management consulting organization based in Maryland, USA.

The second data set used to interrogate the educational attainment variables is the Philippines geodemographic classification of small areas called barangays (villages), which has been exhaustively discussed in Chapter 9 of this book. The geodemographic classification serves as the framework for profiling educational attainment and mapping spatial patterns.

11.3 Methodological Approach

The first step was to extract the data for the NCR from the full DHS data. All participants of recorded in the DHS data set were assigned a geographic coordinate representing their residential location. This made it possible to link the DHS data to the geodemographic classification system as illustrated in Figure 11.1. The maps show the spatial configuration of geodemographic clusters in the NCR, and the sites from which data were collected across the region during the 2003 round of the DHS.

After linking the two data sets, the geodemographic classification system was then used to analyze gendered differences in the rates of educational attainment across the seven neighborhood types. The "unclassified" category was excluded from the analysis.

The geodemographic profiling of educational attainment provided a significantly increased level of neighborhood-level insight that is useful for educational policy and planning. This is further enhanced by the use of penetration ranking reports where neighborhood types are ranked by attainment

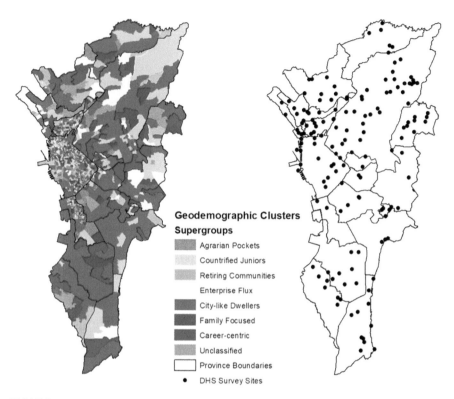

FIGURE 11.1
Geodemographic clusters and DHS data collection sites in the national capital region.

rates, and the cumulative percentage of the type of attainment is compared with the corresponding adult population. The concentration index (CI) (Giorgi and Gigliarano, 2016) has been used in this chapter to illustrate the proportion of inequality in the distribution of attainment in relation to the adult population. The CI is algebraically defined in Equation 11.1 as:

$$CI = \frac{1}{2}\Sigma \left| x_i - y_i \right| \tag{11.1}$$

where:

CI is the index of concentration
x_i is the proportion of adults with a certain level of attainment in geodemographic cluster i
y_i is the proportion of adults in geodemographic cluster i.

Where the index is expressed as a percentage, it ranges from 0 to 100. Where it is not expressed as a percentage, the values are between 0 and 1. A CI value of 0 indicates that there is no inequality, while a value of 1 or 100 means that maximum inequality exists between areas.

11.4 Gendered Differences in Neighborhood Patterns of Educational Attainment

As can be seen from Table 11.1, both males and females tend to be highly educated in the Philippines. However, in comparison with the national adult population, adults based within the NCR seem to be better educated. In terms of the gendered aspects of the analysis, one can immediately see that attainment rates are initially higher for men at primary level, but women catch up as they progress to secondary and higher levels of education.

Results of the geodemographic analysis are presented for males and females in Tables 11.2 and 11.3. Both tables show rates of different types of educational attainment by geodemographic clusters and compare these rates with data at regional levels. The cells shaded in red highlight those geodemographic clusters where rates of attainment are below the regional value.

With regard to primary school attainment, Tables 11.2 and 11.3 show that Agrarian Pockets account for the least attainment rates in comparison to the NCR. Within Agrarian Pockets, there is an above average representation of children and people aged over 65 years. It is also very common to find households where the head has a stepchild. Residents of this type of neighborhood are mostly employed in the agricultural sector. There is also an above average presence of cooperative societies and establishments in the media industry. When compared with the national mean, the incidence of population needing care for disability is quite high.

Although data in the third columns of Tables 11.2 and 11.3 show that primary attainment rates are the lowest in Agrarian Pockets for both genders, evidence indicates that attainment rates for males are approximately half of their female counterparts. Two other neighborhood types (City-like

TABLE 11.1

National and regional levels of educational attainment for males and females (2003)

Level of Attainment	Philippines Men (%)	National Capital Region Men (%)	Philippines Women (%)	National Capital Region Women (%)
No education	1.8	0.2	1.4	0.2
Primary	30.2	13.6	23.1	11.5
Secondary	43	46.1	44.8	45.8
Higher	25	40.1	30.7	42.5
Total	100	100	100	100

Source: Author's calculation based on data from the 2003 Philippines DHS.

TABLE 11.2
Levels of educational attainment for males by geodemographic clusters (2003)

Geodemographic Supergroup	No Education (%)	Primary (%)	Secondary (%)	Higher (%)
Agrarian Pockets	*	3.1	37.4	59.5
Countrified Juniors	*	15.1	52.1	32.8
Retiring Communities	*	14.7	51.5	33.8
Enterprise Flux	*	30.3	39.2	30.5
City-like Dwellers	0.2	11	44.9	43.9
Family Focused	*	8.6	52.9	38.5
Career-centric	0.7	17.3	43.6	38.4
National Capital Region	0.2	13.6	46.1	40.1

Source: Author's calculation based on data from the Philippines geodemographic classification and the 2003 Philippines DHS.

* Suppressed due to no data.

TABLE 11.3
Levels of educational attainment for females by geodemographic clusters (2003)

Geodemographic Supergroup	No Education (%)	Primary (%)	Secondary (%)	Higher (%)
Agrarian Pockets	*	6	40	54
Countrified Juniors	0.5	14.6	45.1	39.8
Retiring Communities	*	17.2	48	34.8
Enterprise Flux	*	12.8	46.5	40.7
City-like Dwellers	0.2	10.6	46.4	42.8
Family Focused	0	9.1	43.3	47.6
Career-centric	0.8	11.8	45.6	41.8
National Capital Region	0.2	11.5	45.8	42.5

Source: Author's calculation based on data from the Philippines geodemographic classification and the 2003 Philippines DHS.

* Suppressed due to no data.

Dwellers and Family Focused) also exhibit below regional average rates of primary school attainment for both genders, indicating some uniformity in the neighborhood-level characteristics associated with low primary school attainment.

When one considers the gendered dimensions of secondary school attainment rates shown in the fourth columns of Tables 11.2 and 11.3, it is evident that there are differences in the neighborhood-level characteristics associated with secondary attainment rates for males and females. For males, the neighborhood types with attainment rates below the regional average include Agrarian Pockets, Enterprise Flux, Career-centric, and City-like Dwellers. For females, the neighborhood types with secondary attainment rates below the regional average are Agrarian Pockets, Family Focused, Countrified Juniors, and Career-centric. Some of the neighborhood characteristics associated with low secondary school attainment rates for males include higher than average population concentration, a dominant age group of 21 to 44 years, and disproportionate presence of Filipino speakers. Conversely, some of the neighborhood characteristics associated with low secondary school attainment rates for females include a dominant presence of children and teenagers aged between 6 and 20 years, and a large number of households whose members are migrant/overseas workers.

With regard to higher level education, it can clearly be seen that males in neighborhood typology called Countrified Juniors have the least attainment rates. A key feature of these neighborhoods is the large concentration of young children and teenagers. This neighborhood type has the largest national proportion of children aged zero to five years. The key employment sector is the agricultural sector, which also includes fishing. Countrified Juniors also have a high presence of small establishments and the many cooperative establishments. The neighborhoods also enjoy a very large presence of recreational establishments and a relatively high concentration of retail trade and media establishments.

The fifth column in Table 11.3 shows the rates of higher education attainment for females compared to the regional average. In contrast with the results for males, females in Retiring Communities account for the lowest higher education attainment rates. This neighborhood type exhibits traits that are in sharp contrast with those associated with low male attainment rates. Households in Retiring Communities are dominated by people of older age categories – over 45 years and nearing the end of their working lives. It is also common to find households with unmarried persons, many of whom are widows. The representation of Filipino speakers is just above the national mean within these neighborhoods. Employment in the educational, public, and defense sectors is quite common. Furthermore, there are also a disproportionately high number of people who work outside the country.

11.5 The Scale of Spatial Inequalities in Educational Attainment

To get a sense of the potential of the analysis presented in this chapter, one might draw upon a claim made in the 2003 DHS report for the Philippines. An extract from that report is presented here.

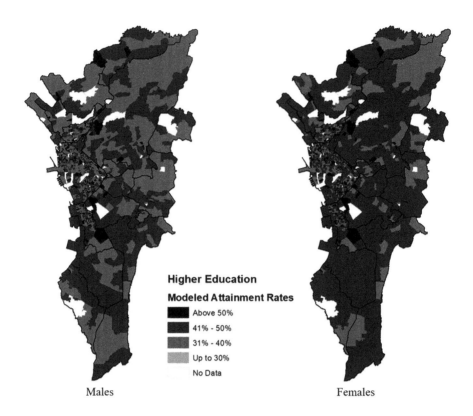

Males   Females

FIGURE 11.2
Map of modeled attainment rates for males and females in the national capital region.

> No major gender differences are observed for education. However, a sig-
> nificant difference is noted between urban and rural areas; the educa-
> tional system favors residents of urban areas.
>
> (NSO and ORC Macro, 2004, p. 11)

The above statement highlights the weakness of current analytical approaches
used for interrogating and reporting the DHS data set. At present, the survey
is analyzed and reported at the regional level, which makes it impossible to
glean insight about what is happening at the local level.

In Figure 11.2, higher education attainment rates for males and females
are extrapolated to the barangay level (i.e., the finest administrative geography) across the NCR. Achieving an estimate of the local spatial variation
in variables such as this (which are often key concerns in many local communities in developing countries) is inherently problematic and can be
quite resource intensive. The modelling of local spatial outcomes through
geodemographic segmentation techniques therefore poses an efficient solution to this difficulty in developing countries.

The task of ensuring geographic and demographic equity in access to educational services in order to increase attainment is a vital obligation for the Philippines government. Increasing and maintaining the focus on equity requires not only focused policymaking and practice, but also a very solid evidence base for closing gaps. The geodemographic analysis of educational attainment reported in Section 11.4 provides a significantly increased level of intelligence for educational stakeholders.

This is further enhanced by the use of ranking reports where geodemographic types are ranked by educational attainment rates, and the cumulative percentage of the type of educational attainment is compared with the corresponding adult population. Tables 11.4 and 11.5 present such reports for male and female primary school attainment by geodemographic supergroups. The tables feature the cumulative proportion of adults with primary level of attainment in the NCR (x_i) and the cumulative proportion of adults in the NCR (y_i). Such reports have been generated for a variety

TABLE 11.4
Ranking report for male primary school attainment in the national capital region (2003)

Geodemographic Supergroup	x_i	y_i	$\|x_i - y_i\|$	Cumulative (x_i)	Cumulative (y_i)
Agrarian Pockets	0.02	0.07	0.05	0.02	0.07
Family Focused	0.04	0.07	0.03	0.06	0.13
City-like Dwellers	0.38	0.46	0.08	0.44	0.59
Retiring Communities	0.08	0.08	0.00	0.52	0.66
Countrified Juniors	0.11	0.10	0.01	0.63	0.76
Career-centric	0.23	0.18	0.05	0.86	0.94
Enterprise Flux	0.14	0.06	0.08	1.00	1.00

Source: Author's calculation based on data from the Philippines geodemographic classification and the 2003 Philippines DHS.

TABLE 11.5
Ranking report for female primary school attainment in the national capital region (2003)

Geodemographic Supergroup	x_i	y_i	$\|x_i - y_i\|$	Cumulative (x_i)	Cumulative (y_i)
Agrarian Pockets	0.04	0.07	0.03	0.04	0.07
Family Focused	0.06	0.08	0.02	0.10	0.15
City-like Dwellers	0.43	0.45	0.02	0.53	0.61
Career-centric	0.17	0.17	0.00	0.70	0.77
Enterprise Flux	0.07	0.06	0.01	0.77	0.84
Countrified Juniors	0.11	0.09	0.02	0.88	0.92
Retiring Communities	0.12	0.08	0.04	1.00	1.00

Source: Author's calculation based on data from the Philippines geodemographic classification and the 2003 Philippines DHS.

TABLE 11.6
Concentration indices for male and female educational attainment in the NCR (2003)

Level of Attainment	Concentration Index for Males	Concentration Index for Females
Primary	0.2	0.11
Secondary	0.6	0.02
Higher	0.9	0.04

Source: Author's calculation based on data from the Philippines geodemographic
classification and the 2003 Philippines DHS.

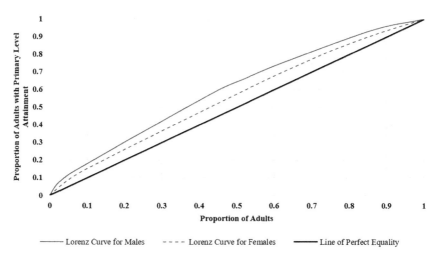

————— Lorenz Curve for Males   — — — Lorenz Curve for Females   ————— Line of Perfect Equality

FIGURE 11.3
Lorenz Curve of primary level attainment versus adult population in the national capital
region (Author's elaboration based on data from the Philippines geodemographic classification
and the 2003 Philippines DHS).

of educational attainment profiles and a measure of inequality (the CI)
estimated.

The concentration indices presented for the NCR in Table 11.6 are quite
instructive. For males, one can discover that geographic inequalities in
attainment expand as one progresses toward up in the educational ladder.
For females, the concentration indices suggest the opposite. One can discover that geographic inequalities among females are the greatest at the primary level. Spatial inequality among females shrinks at the secondary level
of attainment and widens slightly at higher levels of education. This type of
insight can be useful for the design of educational policies, planning, and
programming. Figure 11.3 illustrates the data presented in Table 11.5 as a
Lorenz curve.

11.6 Conclusion

Nobody determines their family, background, or personal characteristics. However, educational attainment is an attribute that can be determined through public policy and individual choices. The educational attainment of an individual can exert significant influence on his or her employment chances, health status, income, and social mobility. Across the world, many countries have come to associate educational attainment with human capital because, by logic, long years of schooling have the potential to translate into a much better skilled workforce, which in turn helps to sharpen productivity. The study described in this chapter examined the gendered dimensions of educational attainment in the NCR of the Philippines using a spatial segmentation framework. The results show that local level disparities in attainment exist for males and females in the region. This chapter has further demonstrated that the deployment and application of spatial analytical principles and techniques, such as those used in this study, can be helpful in interpreting geographically referenced data sets for the educational sector in developing countries and constructing specialized and bespoke solutions.

References

Becker, S., Peters, D.H., Gray, R.H., Gultiano, C. and Black, R.E. (1993). The Determinants of Use of Maternal Child Health Services in Metro Cebu, the Philippines. *Health Transition Review*, 3(1), 77–91.

Caldwell, J.C. (1981). Maternal Education as a Factor in Child Mortality. *World Health Forum*, 2(1), 75–78.

Chiu, M.M. (2007). Families, Economies, Cultures, and Science Achievement in 41 Countries: Country, School, and Student-Level Analyses. *Journal of Family Psychology*, 21(3), 510–519.

Crawford, L. and Sidener, N.L. (1982). *Women's Formal Education and Economic Growth: The Case of the Philippines*. Paper presented at the 5th National Conference on the Third World, Omaha, Nebraska, 27–30 October, 1982.

Damon, A., Glewwe, P., Wisniewski, S. and Sun, B. (2016). *Education in Developing Countries – What Policies and Programmes Affect Learning and Time in School?* Stockholm: The Expert Group for Aid Studies.

Dorling, D. and Woodward, R. (1996). Social Polarization 1971–1991: A Micro-Geographical Analysis of Britain. *Progress in Planning*, 45(2), 63–122.

Dyer, C. (2010). Education and Social (In)justice for Mobile Groups: Re-Framing Rights and Educational Inclusion for Indian Pastoralist Children. *Educational Review*, 62(3), 301–313.

Giorgi, G.M. and Gigliarano, C. (2016). The Gini Concentration Index: A Review of the Inference Literature. *Journal of Economic Surveys*, 31(4), 1130–1148.

Grant, M.J. and Behrman, J.R. (2010). Gender Gaps in Educational Attainment in Less Developed Countries. *Population and Development Review*, 36(1), 71–89.

Halle, T.G., Kurtz-Costes, B. and Mahoney, J.L. (1997). Family Influences on School Achievement in Low-Income, African American Children. *Journal of Educational Psychology*, 89(3), 527–537.

King, E.M. and Hill, M.A. (1993). *Women's Education in Developing Countries: Barriers, Benefits and Policies.* Washington, DC: The World Bank.

Lewin, K.M. (2015). *Educational Access, Equity, and Development: Planning to Make Rights Realities.* Paris: The United Nations Educational, Scientific and Cultural Organization.

Mertaugh, M.T., Jimenez, E.Y. and Patrinos, H.A. (2009). *The Global Challenge in Basic Education: Why Continued Investment in Basic Education is Important.* Washington, DC: The World Bank.

NSO and ORC Macro (2004). *National Demographic and Health Survey 2003.* Calverton, MD: NSO/Philippines and ORC Macro.

Tilbury, D., Stevenson, R.B., Fien, J. and Schreuder, D. (Eds). (2002). *Education and Sustainability: Responding to the Global Challenge.* Cambridge: International Union for Conservation of Nature and Natural Resources Publications Services Unit.

UIS (2020). *Philippines.* Montreal: UNESCO Institute of Statistics.

UN (1998). *Principles and Recommendations for Population and Housing Censuses, Revision 1.* New York, NY: United Nations.

12

Conclusion

12.1 Geodemographics for the Developing World

> For, indeed any city, however small, is in fact divided into two, one the
> city of the poor [and] the other of the rich; these are at war with one
> another; and in either there are many smaller divisions, and you would
> be altogether beside the mark if you treated them as a single State. But if
> you deal with them as many and give the wealth or power or persons of
> the one to the other, you will always have a great many friends and not
> many enemies.
>
> (Knight and Jowett, 1999, pp. 137–138).

In the extract above, Plato was referring to one of the proposals of Socrates
about how inequality should be tackled in ancient Greece. The policy solution offered by Socrates at the time was to embark on a targeting strategy that
differentiated people and places based on their characteristics rather than on
a one-size-fits-all approach.

More than 2000 years after Socrates offered these proposals, Charles Booth,
a social reformer in London, began an inquiry aimed at debunking poverty-
related findings of the Socialist Federation (Pfautz, 1967). Booth's work was
novel in many respects. First, it allowed for the production of street-level
maps of social conditions of people living in London. The categorizations
and color coding of these maps were based on the premise that social heterogeneity occurs within spatial structures. Booth's approach to conducting
the analysis served as a useful foundation for contemporary and modern-
day multivariate spatial analysis (Pfautz, 1967). Many researchers converged
and agreed that modern-day development and deployment of small area
classifications originated from the work of Booth (Harris et al., 2005; Burrows
and Gane, 2006; Vickers, 2006).

Following Booth's experiments, the concept of social area classification extended into the fields of urban sociology, linking up with multiple

urban sociological models that are associated with the Chicago School (Robson, 1971).

The idea of geodemographics is underpinned by the notion that people and places that are close to each other are likely to exhibit similar characteristics (Tobler, 1970; Harris et al., 2005). Well and beyond the notion that closely located people or places are alike, geodemographic intricacies take the laws of spatial dependence a step further to infer that geographic proximity is not necessarily a perfect criterion for estimating similarity as distant features may also have characteristics that are very similar to each other (Vickers and Rees, 2006).

Not all areas have the same needs (Ballas et al., 2005); however, all areas within a defined geographic boundary can be summarized to a limited number of types (Miller and Han, 2001). Therefore, knowledge of the types of areas that exist in space from the types of people living in them can be very useful in ascertaining the types and levels of needs that exist.

Some developed countries have benefited from understanding the spatial dimensions of local level disparities in the social makeup of their societies (Brown, 1991; Brown et al., 2000). This has helped to reshape what Dorling and Ballas (2008) described as a pear-shaped nature of poverty and inequality in some of these countries. Tackling social and spatial inequality at the roots, and at local geographic scales, is therefore very important for bridging the gaps between members of human societies. Consequently, small area geodemographic classifications are being deployed within public sector and academic research because of their ability to provide useful and effective evidence-based summaries at local spatial scales. However, developing countries are still faced with the challenge of tackling and targeting human and social problems with solutions at small spatial scales, and sustaining good-practice solutions at the same scale (Ojo and Ezepue, 2011).

At a well-received talk titled "Let my dataset change your mindset" delivered by the late Professor Hans Rosling during the summer of 2009 to the US State Department, Professor Rosling challenged several public policy practices in developing countries. Rosling (2009) explained how many of the world's global policymakers are uninformed of global social trends. He painted a powerful picture of global convergence in his talk. Of greater importance, however, was the fact that he stressed the need for global development agencies to move away from current approaches of social and spatial analysis which mask everyday lived experiences of people at the grass roots. Rosling called for a discarding of simplistic extrapolation of data noting that if the problem of inequality is to be addressed, then it has to be made a local issue.

This book is offered as a tool for addressing some of the important issues raised by Rosling and many others like him. Policy stakeholders across developing countries need effective decision-making tools that could help them conduct analysis, display and disseminate results, and make informed decisions that have direct and indirect implications on grassroots populations. Small area geodemographic classifications are proven to be powerful and effective in delivering these functionalities to help economic developers sustain economic recovery and growth. Geodemographics brings together technologies, disciplines, and practical techniques such as Geographic Information Science, Data Science, Machine Learning, Social Sciences, Environmental Sciences, and Development Studies. Furthermore, geodemographic principles and practice can be integrated into multiple sectors of the economy. Embracing geodemographics should also allow governments in developing countries to consider different economic sectors as an important part of an ecosystem, reducing inequalities, supporting jobs, and potentially generating revenue for government.

12.2 Some Lessons to Take Home

This book subsumes multiple theoretical and practical lessons that the author hopes readers will bear in mind going forward. Some of these lessons are summarized here for ease of comprehension. It should be noted that the narratives presented here are summaries of the big picture contained in the book.

The purpose of any small area classification is quite important, and it should be defined as clearly as possible at the outset of developing these classifications. A clear definition of purpose will help guide the choice of input data and building blocks of the classification. For developing countries, a clear definition of purpose can act as a driving force for connecting with relevant stakeholders and potential users of an area classification.

Another important lesson that the readers should bear in mind is the importance of policy relevance when developing geodemographic classifications for developing countries. It is important for researchers to demonstrate how their classifications can improve the policymaking landscape for public good if they wish to secure buy-in from relevant stakeholders. Doing this can also ease difficulties associated with accessing data from government agencies. Furthermore, small area classifications that can demonstrate relevance to public policy are probably more likely to attract the attention of voluntary sector organizations.

Organizational attitudes of statistical agencies in developing countries are influenced by cultural characteristics. It is therefore important for academic researchers and practitioners to understand these organizational characteristics before making contact with relevant agencies. This book addresses organizational culture in Chapter 4. Primary evidence indicates that cultural practices constitute barriers to accessing data. However, some of these barriers can be broken down if researchers make painstaking efforts to acquire knowledge about the culture of organizations in developing countries.

12.3 Issues with Crowdsourcing Data

Researchers in developing countries are increasingly considering crowdsourced data for spatial analysis. The process of crowdsourcing data entails making an open call to members of the public to engage in some form of consultation exercise. Such consultations may be accompanied by forms of compensation to encourage participation. While there are some benefits of doing this, it is also important to make a note of caution.

The proliferation of ICTs and the penetration of mobile phones across developing countries (Gunasekarana and Harmantzis, 2007) provide unique opportunities for conducting rapid public consultations. Social media spaces like social networking sites (Facebook, Twitter, and Instagram) also offer researchers an avenue to recruit "crowds" for sourcing sociodemographic data that may be relevant to national and local policy issues (Singleton and Longley, 2009). Most national surveys supported by international donor agencies are conducted on a face-to-face basis, but gathering face-to-face data in developing countries is an expensive activity. Furthermore, it is time-consuming and can sometimes generate small sample sizes. The greatest benefit of crowdsourcing data for spatial analysis in developing countries is that the procedure can help increase samples in a relatively cost-effective manner.

However, despite the rapid proliferation of ICT in developing countries, there are still substantial numbers of people without access to these technologies. For instance, although three-quarters of the population in sub-Saharan Africa had a subscriber identification module (SIM) connection in 2018, mobile subscriber penetration was just 44%, well behind the global average of 66% (GSMA, 2019). The share of unique mobile subscribers in the region is predicted to rise to 50% by 2025.

Access to ICT in rural areas of developing countries is still a huge challenge (Chapman and Slaymaker, 2002). In order to truly take advantage of crowdsourcing, one needs a large diverse crowd. Ideally, contributors of crowdsourced data should be broadly spread across geographic areas and socioeconomic groups in order to avoid unrepresentative samples. The digital divide between most urban and rural areas is remarkably high in the developing countries (Bjørn and Stein, 2007), which makes it challenging to acquire geographically representative data through existing ICT platforms.

In addition to substantial urban-rural population inequalities in access to ICTs, another problem that can arise from crowdsourcing data is that "noise" may be introduced into the data in the form of multiple responses from the "crowd." Incentives are sometimes used to encourage the public participation in crowdsourcing (IRMA, 2019). Sometimes, members of the crowd may provide multiple responses, which can skew the overall data set. This is quite common where incentives like entries into raffle draws are used to encourage participation in crowdsourcing. Participants may be tempted to increase their chances of winning the draw by providing multiple responses using different identities. To ameliorate this problem, smart computing techniques for Internet Protocol (IP) address monitoring can be used to discourage multiple entries. However, this type of deterrent may restrict survey response to one entry per computer. In some developing countries where most households do not own a personal computer and access to the Internet is restricted, (Bjørn and Stein, 2007), the use of IP address monitoring can have knock-on effects on sample sizes.

In addition to the problems of data collection errors, poor Internet speed may discourage participation in crowdsourcing activities. High-speed Internet provision is not widespread across developing countries (GSMA, 2019) and this can frustrate participation in online data crowdsourcing.

In most developing countries, the degree of current challenges facing most crowdsourcing methods is still quite significant. Urban–rural divides in the proliferation of ICTs can be disadvantageous for rural dwellers, leading to underrepresentations. Furthermore, crowdsourcing data in developing countries have been shown to favor younger aged citizens (Verhulst and Young, 2017).

12.4 Coupling Official Face-to-Face Surveys with Emerging Forms of Data

Given the significant challenges associated with using crowdsourced data alone for the development of small area geodemographic classifications, it is impossible to neglect the importance of official face-to-face data collection. The conclusion of this book is that the proliferation of geodemographic classifications in developing countries can be enhanced by combining official surveys collected via face-to-face methods with new and emerging forms of data.

Developing countries are inextricably linked to the rest of the world and it is recognized that we are living in an information age. Consequently, extraordinary volumes of data are constantly being generated as a result of the increasing pervasiveness of technology and connectivity. New and emerging forms of technology are providing options for rapid data generation. There is no consensus in the literature as to how to classify the themes of emerging technologies that are shaping public policy delivery in developing countries. However, some sources of new and emerging forms of data (OECD, 2013) that can be used to complement traditional surveys and for understanding human conditions are listed here. It is important to mention that the list is not exhaustive.

- Online search terms
- Website interactions

- Social media (text-based and multimedia)
- Blogs, forums, and news sites
- Closed-circuit television (CCTV)
- Sensors and connected devices
- Mobile phone data
- GPS tracking data
- Satellite/aerial images
- Nighttime visible radiation.

The conversion of these emerging forms of data into actionable information requires the use of advanced computational techniques to unveil trends and patterns within and between these extremely large socioeconomic data sets. New insights derived from mining these data sets could complement official survey statistics in developing countries.

References

Ballas D., Rossiter D., Thomas B., Clarke G. and Dorling D. (2005). *Geography Matters. Simulating the Local Impacts of National Social Policies.* York: Joseph Rowntree Foundation.

Bjørn, F. and Stein, K. (2007). A Rural-Urban Digital Divide? Regional Aspects of Internet Use in Tanzania. *Proceedings of the 9th International Conference on Social Implications of Computers in Developing Countries, São Paulo, Brazil*, 28–30 May.

Brown, P.J. (1991). Exploring Geodemographics. In: I. Masser and M.J. Blakemore (Eds), *Handling Geographical Information.* London: Longman.

Brown, P.J.B., Hirschfield, A.F.G. and Batey, P.W.J. (2000). *Adding Value to Census Data: Public Sector Applications of Super Profiles Geodemographic Typology.* Working Paper 56, URPERRL, Department of Civic Design, University of Liverpool.

Burrows, R. and Gane, N. (2006). Geodemographics, Software and Class. *Sociology,* 40(5), 793–812.

Chapman, R. and Slaymaker, T. (2002). *ICTs and Rural Development: Review of the Literature, Current Interventions and Opportunities for Action.* London: Overseas Development Institute.

Dorling, D. and Ballas, D. (2008). Spatial Divisions of Poverty and Wealth. In: T. Ridge and S. Wright (Eds), *Understanding Poverty, Wealth and Inequality: Policies and Prospects.* Bristol: Policy Press.

GSMA (2019). *The Mobile Economy Sub-Saharan Africa 2019.* London: GSM Association.

Gunasekarana, V. and Harmantzis, F.C. (2007). Emerging Wireless Technologies for Developing Countries. *Technology in Society,* 29(1), 23–42.

Harris, R., Sleight, P. and Webber, R. (2005). *Geodemographics, GIS and Neighborhood Targeting.* London: Wiley.

IRMA (2019). *Smart Cities and Smart Spaces: Concepts, Methodologies, Tools, and Applications: Concepts, Methodologies, Tools, and Applications*. Hershey, PA: IGI Global.

Knight, M.J. and Jowett, B. (1999). *The Essential Plato*. New York, NY: Quality Paperback Book Club.

Miller, H.J. and Han, J. (2001). *Geographic Data Mining and Knowledge Discovery*. New York, NY: Taylor & Francis.

OECD (2013). *Data for Understanding the Human Condition*. Paris: The Organization for Economic Co-operation and Development.

Ojo, A. and Ezepue, P. O. (2011). How Developing Countries Can Derive Value from the Principles and Practice of Geodemographics, and Provide Fresh Solutions to Millennium Development Challenges. *Journal of Geography and Regional Planning*, 4(9), 505–512.

Pfautz, H.W. (1967). Sociologist of the City. In: C. Booth (Ed), *On the City: Physical Pattern and Social Structure* (selected writings). Chicago, IL: University of Chicago Press.

Robson, B.T. (1971). *Urban Analysis: A Study of City Structure*. Cambridge: Cambridge University Press.

Rosling, H. (2009). *Let My Dataset Change Your Mindset*. Available at: www.ted.com/talks/hans_rosling_at_state.html. Accessed on 14 January 2020.

Singleton, A.D. and Longley, P.A. (2009). Geodemographics, Visualization, and Social Networks in Applied Geography. *Applied Geography*, 29(3), 289–298.

Tobler, W.R. (1970). A Computer Movie Simulating Urban Growth in the Detroit Region. *Economic Geography*, 46, 234–240.

Verhulst, S.G. and Young, A. (2017). *Open Data in Developing Economies: Toward Building an Evidence Base on What Works and How*. Cape Town: African Minds.

Vickers, D.W. (2006). *Multi-Level Integrated Classifications Based on the 2001 Census*. Unpublished PhD thesis. School of Geography, University of Leeds.

Vickers, D. and Rees, P. (2006). Introducing the Area Classification of Output Areas. *Population Trends*, 125, 15–24.

Index

9780367652326